Process Development Symposium 2016

AA009105

Development Challenges in Sustainable Processing

Providence, Rhode Island, USA
7-9 June 2016

ISBN: 978-1-5108-2576-5

Printed from e-media with permission by:

Curran Associates, Inc.
57 Morehouse Lane
Red Hook, NY 12571

Some format issues inherent in the e-media version may also appear in this print version.

Printed by Curran Associates, Inc. (2016)

AIChE
120 Wall Street, FL 23
New York, NY 10005-4020

Phone: (800) 242-4363
Fax: (203) 775-5177

www.aiche.org

Additional copies of this publication are available from:

Curran Associates, Inc.
57 Morehouse Lane
Red Hook, NY 12571 USA
Phone: 845-758-0400
Fax: 845-758-2633
Email: curran@proceedings.com
Web: www.proceedings.com

Process Development Symposium 2016

Development Challenges in Sustainable Processing

Providence, Rhode Island, USA
7-9 June 2016

AIChE
The Global Home of Chemical Engineers

TABLE OF CONTENTS

KEYNOTE

COMMON DEVELOPMENT AND SCALE-UP CHALLENGES FOR HIGH WATER AND ENERGY USE PROCESSES

ALTERNATIVE PROCESS TECHNOLOGIES FOR HIGH-WATER APPLICATIONS

RISK MITIGATION IN PROCESS DEVELOPMENT

SCALE-UP ISSUES INVOLVING DILUTE AQUEOUS STREAMS

PILOTING THERMOCHEMICAL AND FERMENTATION PROCESSES

COUPLING PROCESS DEVELOPMENT WITH LIFE CYCLE ANALYSIS

PROCESS MODELING, OPTIMIZATION, AND ECONOMICS EVALUATION

Advancing a Water Circular Economy

Key to sustainable process development

Tracy Young
Core R&D Director
Consumer & Infrastructure Solutions
The Dow Chemical Company

PDS Process Development Symposium

2016 Process Development Symposium | Providence, RI| June 7-9, 2016

DOW

Tomorrow's Needs Will Be Even Greater

By 2030, we will need
- 30% more water
- 40% more energy
- 50% more food

*Source: World Business Council on Sustainable Development

Water as a Risk for Industry Growth

Water demand for manufacturing is projected to increase 400%*

**Source: U.N., based on projections for 2000-2050. Photo source: Poland Radio*

Water as a Risk for Industry Growth

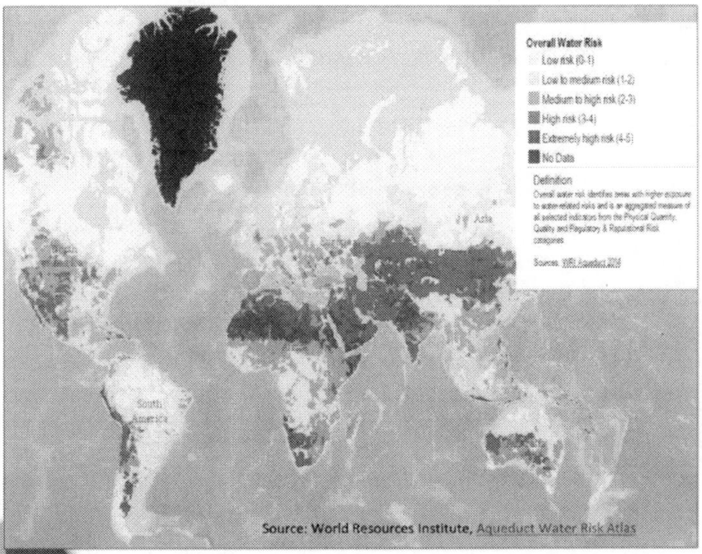

Water Consumption

Municipal – 8%

Agricultural – 70%

Industrial – 22%

Companies are increasingly identifying water as a key risk factor to their supply chain and operation reliability

Water Concerns fall into two categories

Quantity
Ensuring adequate supply

Quality
Meeting water specifications

- Where to build plants so that they have long-term access to water?
- What actions to take when there is unsustainable tapping of aquifers?
- How to reduce water-intensity in manufacturing process?
- How to manage operational and reputational risks in light of competition with other users?

- Does the source water meet the health specifications for product ingredients?
- How to economically and environmentally address brine management?
- Does the salt and metal content pose threat to the integrity of the plant?
- How to properly treat effluent without impacting surrounding environment?

Challenges with quantity and quality are intricately linked

Increased drought or excessive pumping of ground water leads to higher salinity concentrations. →

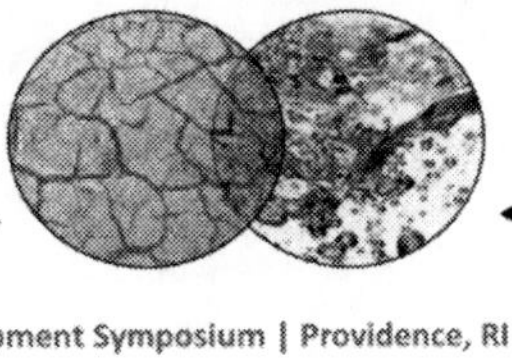

← Anthropogenic or naturally-contaminated water decreases total availability of water.

Step-phase approach to mitigate water risk

- Step 5: Exploring options to mitigate risk
- Step 4: Assessing impact of water risk on the bottom line
- Step 3: Awareness of quality requirements
- Step 2: Evaluating long-term supply of water
- Step 1: Understanding the role of water in the business

Industrial segment as a collective unit is still in the early stages

Corporate water assessments tool examples

Tools target 4 key target areas

- Promote operational efficiency and sustainable product design
- Identify and assess water-related business risk
- Manage water-related impacts on communities and the environment
- Communicate water information to stakeholders

Ceres | WWF | GrowingBlue (Water. Economics. Life.) | World Business Council for Sustainable Development | Global watertool© | VEOLIA WATER | GEMI® Local Water Tool™ (LWT) | WATER RISK MONETIZER (Informing better business decisions) | AQUEDUCT

True Cost/Value of Water Assessment

Cost Related to Risk:

- **Temporary loss of license to operate**
- **Plant shutdown due to water shortage**
- **Penalties due to water pollution**
- **Strengthening of discharge regulation**

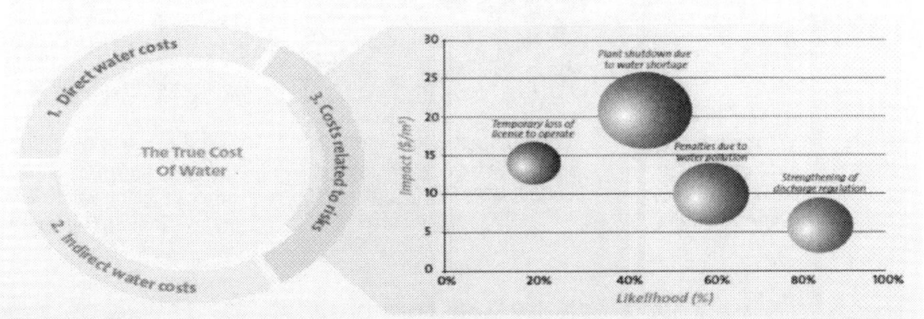

The graph represents an example of identified risks during the analysis. Each risk is plotted on a graph based on its probability and potential economic impact.

Source: Veolia True Cost Water Assessment Tool

Process Development Symposium

Circular Economy

Key to sustainable process solutions for energy/water

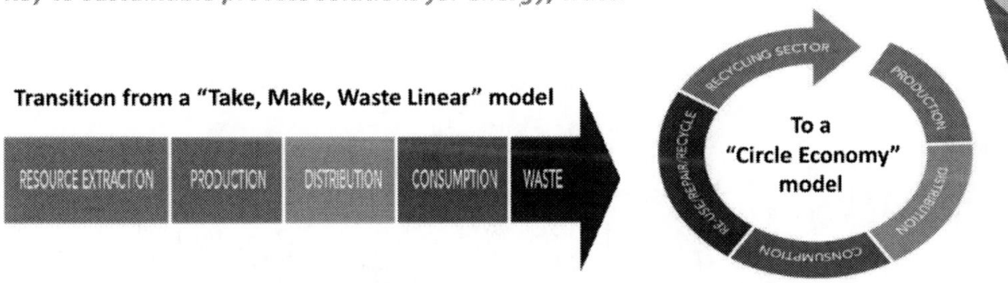

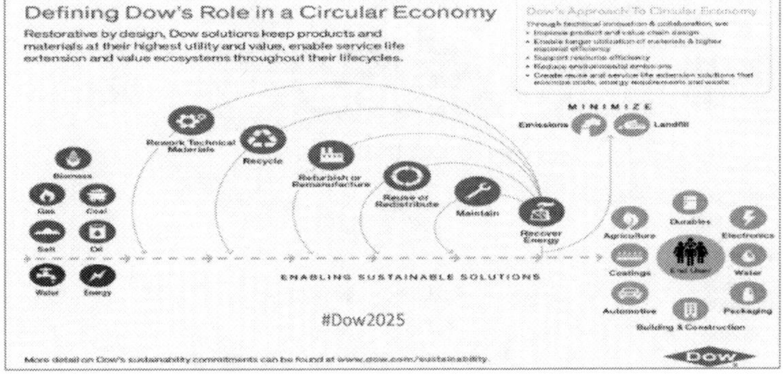

ELLEN MACARTHUR FOUNDATION

Process Development Symposium

Advancing a Circular Economy

Our Aspiration for Water

Public-private partnerships, water policy, and technology to advance breakthroughs in energy and water solutions

Provide treatment solutions to reduce fresh water intake, and enable new sources of clean, potable water.

Close the loop on water, expanding recycled water, reuse of other's reclaim wastewater, and renew alternate sources for reuse

2025 Sustainability Goal commitments at key water stressed sites & partner customers to adopt new water treatment solutions.

Water reuse schemes align Circular Economy

point-of-source	on-site	discharge
RE-CLAIM	RE-CYCLE	RE-NEW
Source reclaimed water	**Increase number of cycles**	**Upgrade for alternate use**

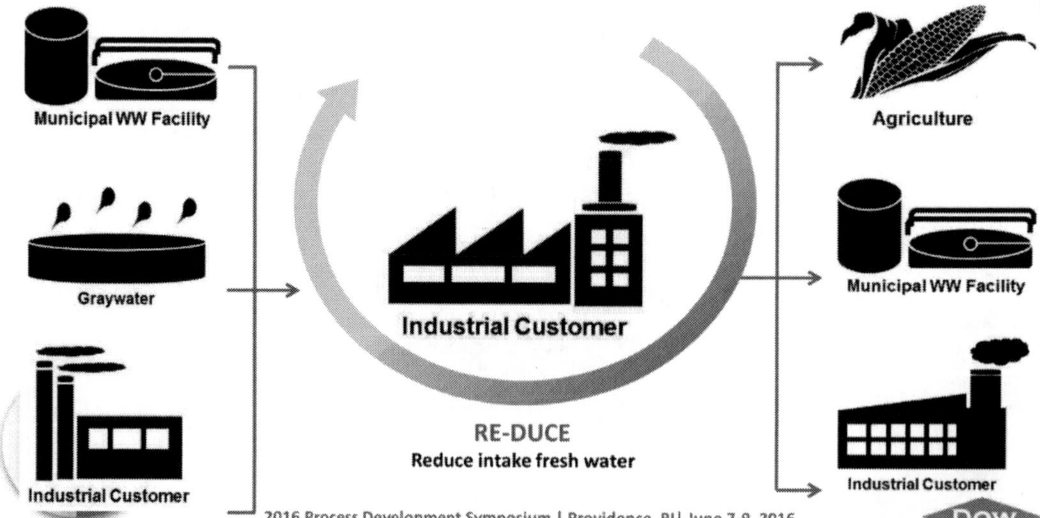

2016 Process Development Symposium | Providence, RI| June 7-9, 2016

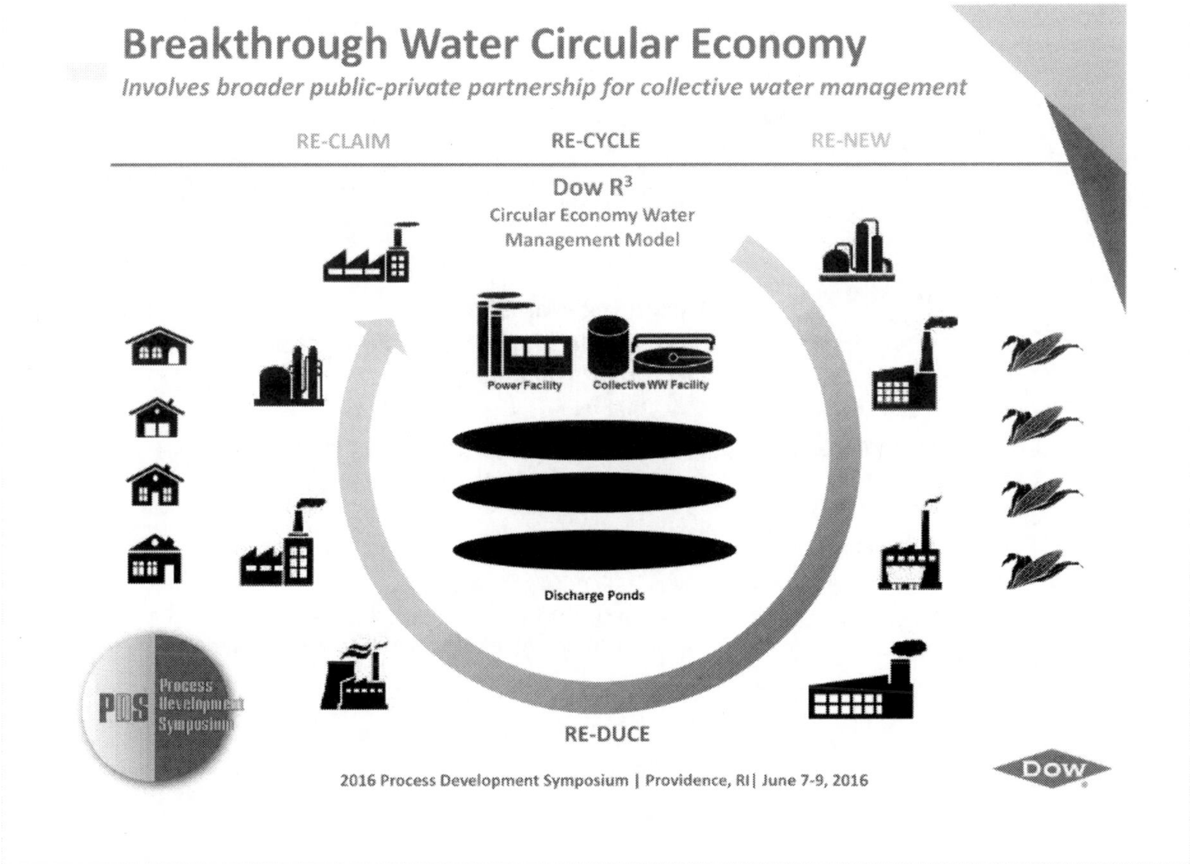
Breakthrough Water Circular Economy
Involves broader public-private partnership for collective water management
RE-CLAIM
RE-CYCLE
RE-NEW
Dow R³
Circular Economy Water
Management Model
Power Facility
Collective WW Facility
Discharge Ponds
RE-DUCE
PDS Process Development Symposium
2016 Process Development Symposium | Providence, RI| June 7-9, 2016
DOW

Dow Terneuzen Netherlands
Advancing a Water Circular Economy to enable growth in the region
DOW
Municipal
Wastewater
reuse 3X
Circular Economy Solutions
for a Scarce fresh water region
Meet Dow's demand:
22 Mm³/yr of fresh water
• Breakthrough Public-Private Partnership
established to accept Municipal Wastewater
•Water re-used first to generate
High Pressure Steam for site
• then reused again for Cooling Towers
• then finally Evaporated
FRESHWATER
scarce
REGION
120 km
CO₂ emissions
60,000 tons/year
Cost
Energy
95% reduction

International Society for Water Solutions

AIChE ISWS Mission

To tackle critical water related issues by presenting and improving potential solutions, optimizing water use while minimizing pollution, collaborating to raise awareness, and sharing knowledge globally.

Upcoming Workshops:
Sustainable Water Management for Mining
Oct 24-26, Denver CO

Interested to join AIChE ISWS Division?
Visit www.aiche.org/ifs/community/isws

2016 Managing Board Chairs
Kim Ogden, University of Arizona
Tracy Young, The Dow Chemical Company
Darlene Schuster, AIChE Executive Director (sponsor)

Thank you

2016 Process Development Symposium | Providence, RI| June 7-9, 2016

Water technology toolbox

Reverse osmosis for desalinating seawater and brackish water

Ultrafiltration for reliable, High-quality water

Fine particle filters that safely recover wastewater for reuse

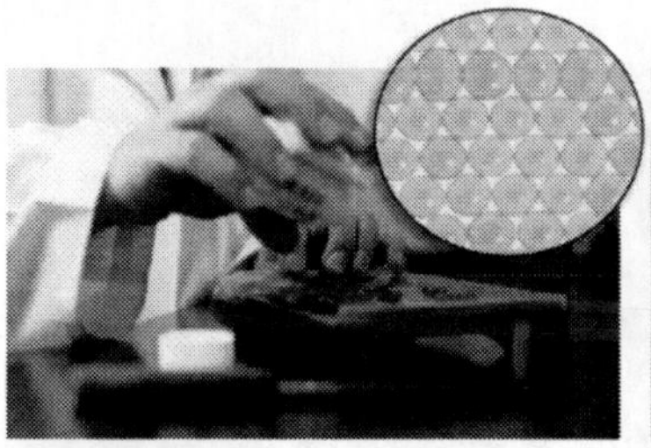

Ion exchange resins for ultrapure water used in manufacturing

Resins that contribute to fewer leaks in irrigation pipes

Pipelines for more reliable delivery of water

the energy of innovation™

Water and Energy Use in the Ethanol Industry

June 8,2016

Douglas B. Rivers, Ph.D.
ICM, Inc.
Director, Technology Relations & Innovation
Direct Line: 316.977.8502

Outline

- ICM Philosophy
- History of Energy and Water Use
- Current Practices
- Opportunities for Improvement
- Conclusions

the energy of innovation™

ICM Record

> Based in Colwich, KS and St. Joseph, MO with ~ 400 Employees
> Technology & Process Integrators
> R&D, Engineering, Design, Manufacturing, Construction
> 102 ICM Designed US Ethanol Plants + 19 Retrofitted US Ethanol Plants
> 9+ Billion Gallons/Year Ethanol
> 24+ Million Tons Animal Feed/Year
> 1.6+ Billion Pounds of Distillers Oil

the energy of innovation™

What is Possible?

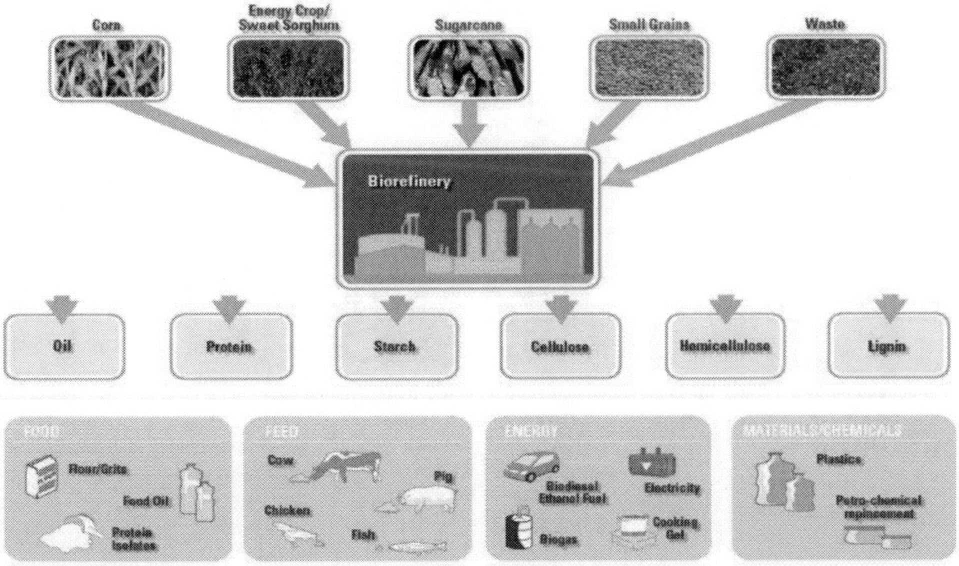

the energy of innovation™

Vertical Value Chain

- The Vertical Value Chain
- Process Trade offs
  - Feedstock Cost
  - Operational Cost
  - Capital Cost
- Major Process Steps
  - Feedstock Handling
  - Feedstock Preparation
  - Conversion
  - Product Separation
  - Product Refinement
  - Storage

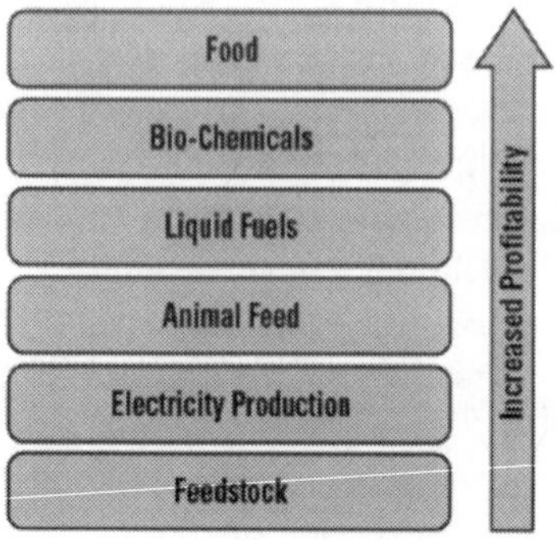

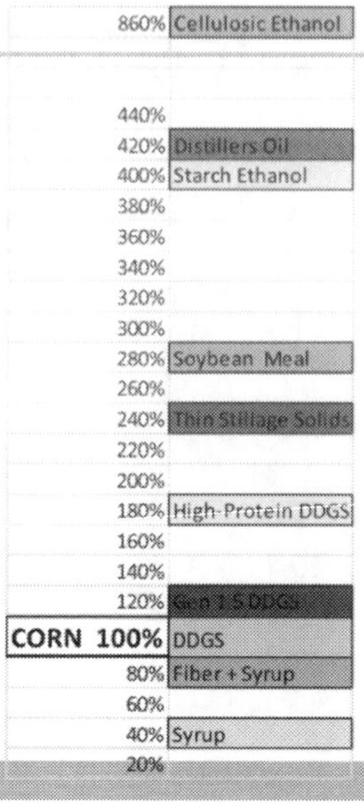

ICM the energy of innovation™

Gen 2.0 Co-Located Bio-Refinery Value Per Acre (1 Corn Acre + 1 Cellulosic Acre)/2

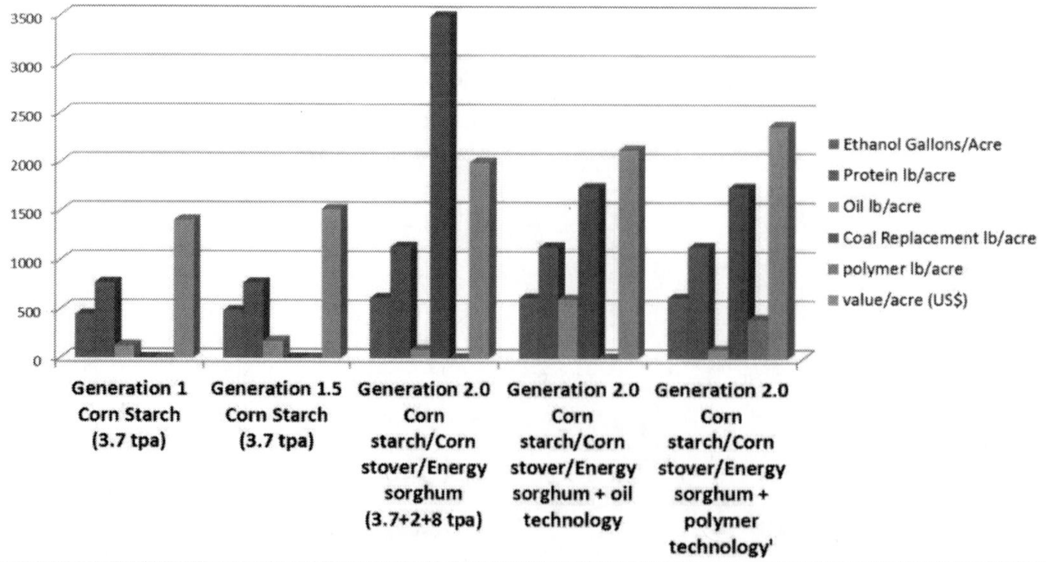

ICM the energy of innovation™

Design Philosophy

- Platform based technologies
- Keep design simple and robust
- Design for high, verifiable conversion yields
- Reduced operating cost
- Ability to react to market conditions

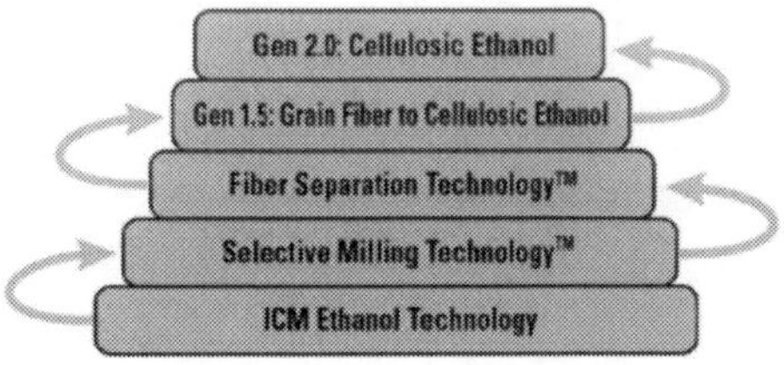

Historical Water Use

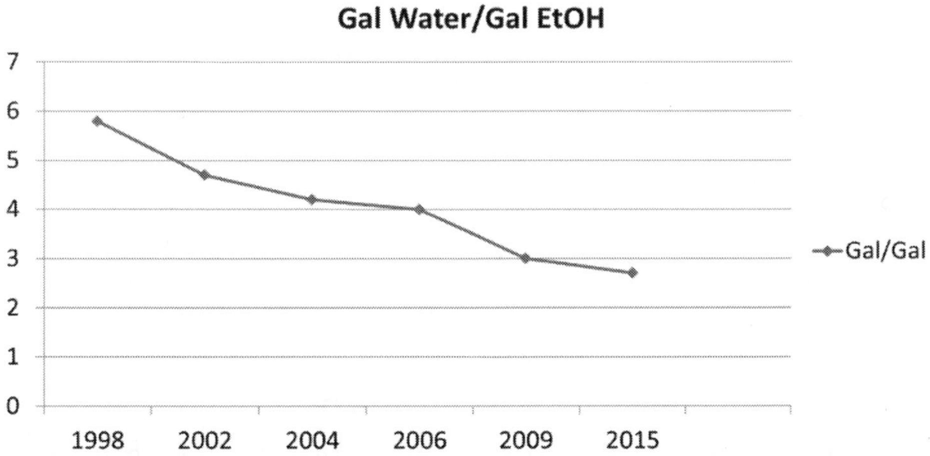

Historical Energy Use

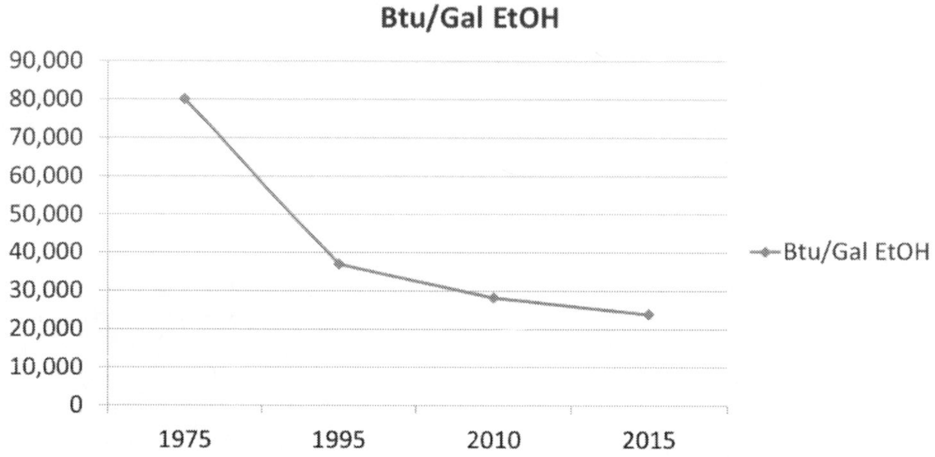

the energy of innovation™

Historical Ethanol Yield

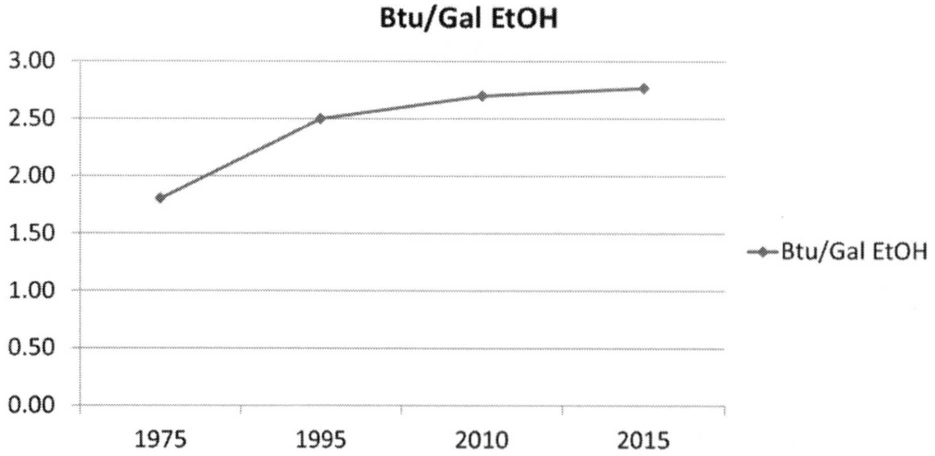

the energy of innovation™

ICM, Inc. Proprietary/Confidential Commercial and Trade Secret Information protected by Contractual Secrecy Agreement. ICM Trade Secret Information protected by U.S. Economic Espionage Act and Uniform Trade Secrets Act. Patents Pending for Selective Milling Technology™, Fiber Separation Technology™ and Generation 1.5 Grain Fiber to Cellulosic Ethanol Technology™. Copying and Distribution Prohibited. © 2016 ICM, Inc. All Rights Reserved.

ICM Philosophy of Water and Energy Use

- Minimize the use of Water and Energy/Product
- Greater Return on Investment to Plant Operations
- Zero Process Water Discharge Design
- Most Aggressive Guarantees in Ethanol Industry

ICM the energy of innovation™

ICM Performance vs Ethanol Industry

- Water Use – 2.7 gal/gal EtOH
  - Industry Average – 3.0 gal/gal EtOH
- Energy Use -
  - 18,500 Btu/gal EtOH
    - Industry Average – 23,900 Btu/gal EtOH
  - 0.60 kWh/gal EtOH
    - Industry Average – 0.7 kWh/gal EtOH
- Ethanol Production – 2.8+ gal/Bu
  - Industry Average – 2.7+ gal/Bu

ICM the energy of innovation™

ICM, Inc. Proprietary/Confidential Commercial and Trade Secret Information protected by Contractual Secrecy Agreement. ICM Trade Secret Information protected by U.S. Economic Espionage Act and Uniform Trade Secrets Act. Patents Pending for Selective Milling Technology™, Fiber Separation Technology™ and Generation 1.5 Grain Fiber to Cellulosic Ethanol Technology™. Copying and Distribution Prohibited. © 2016 ICM, Inc. All Rights Reserved.

What Is The Future?

- Gen 1.0
  - Starch
  - The low hanging fruit is gone
- Gen 1.5
  - Starch + Cellulose
  - Increased EtOH/Bu
- Gen 2.0
  - Cellulose
  - The promise of more from less?

ICM the energy of innovation™

What are the Challenges?

- Gen 1.0
  - Baseline
- Gen 1.0 w/SMT
  - Increased EtOH and Oil
- Gen 1.0 w/FST
  - Increased throughput
- Gen 1.5
  - EtOH and Oil Yield Increase
  - No Chemical Increase
- Gen 2.0 Co-Location
  - Zero Process Water Discharge

ICM the energy of innovation™

ICM, Inc. Proprietary/Confidential Commercial and Trade Secret Information protected by Contractual Secrecy Agreement. ICM Trade Secret Information protected by U.S. Economic Espionage Act and Uniform Trade Secrets Act. Patents Pending for Selective Milling Technology™, Fiber Separation Technology™ and Generation 1.5 Grain Fiber to Cellulosic Ethanol Technology™. Copying and Distribution Prohibited. © 2016 ICM, Inc. All Rights Reserved.

Generation 1.0 Grain to Ethanol™

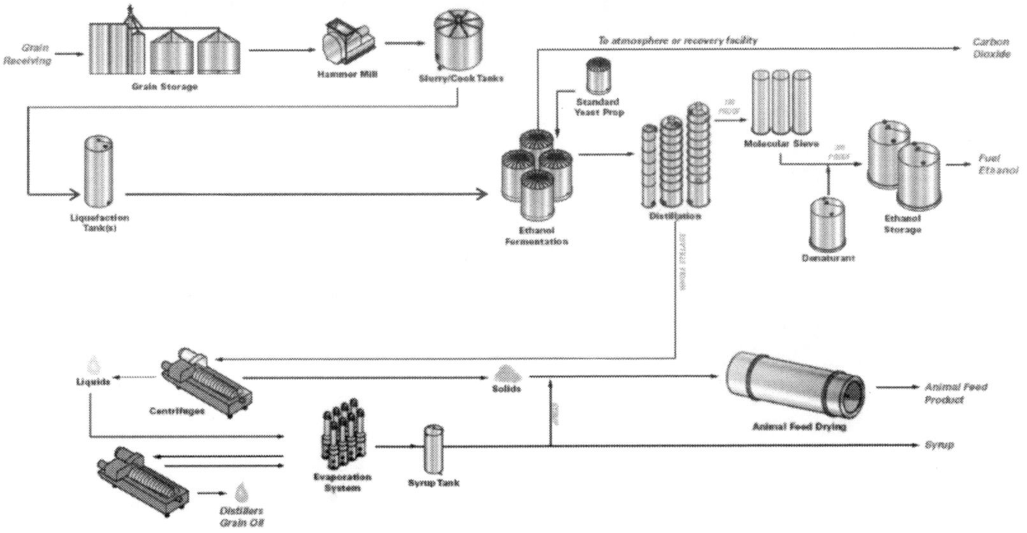

the energy of innovation™

Can Generation 1.0 Grain to Ethanol™ Meet LCFS Standards?

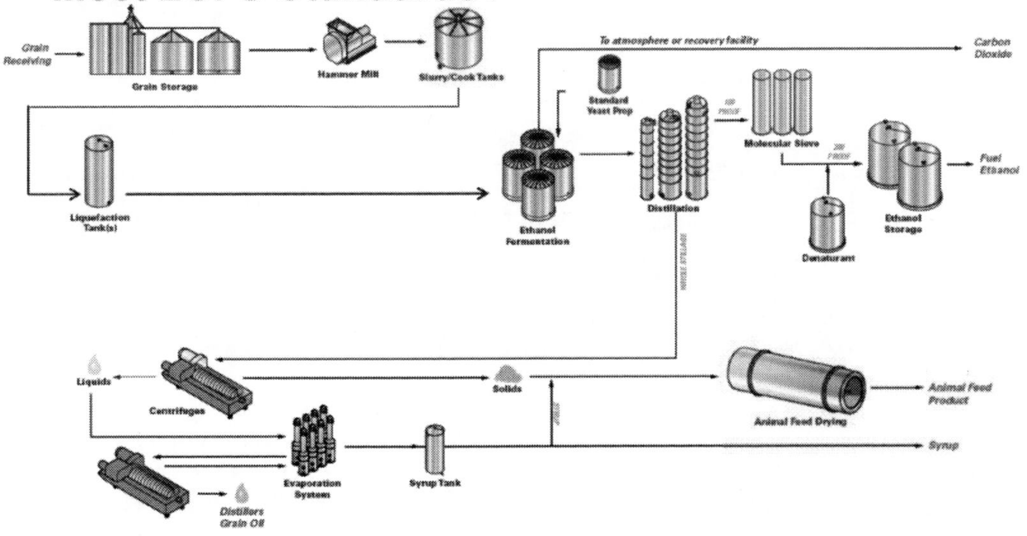

the energy of innovation™

ICM, Inc. Proprietary/Confidential Commercial and Trade Secret Information protected by Contractual Secrecy Agreement. ICM Trade Secret Information protected by U.S. Economic Espionage Act and Uniform Trade Secrets Act. Patents Pending for Selective Milling Technology™, Fiber Separation Technology™ and Generation 1.5 Grain Fiber to Cellulosic Ethanol Technology™. Copying and Distribution Prohibited. © 2016 ICM, Inc. All Rights Reserved.

Generation 1.5 Grain Fiber to Ethanol™ with SMT™ and FST™

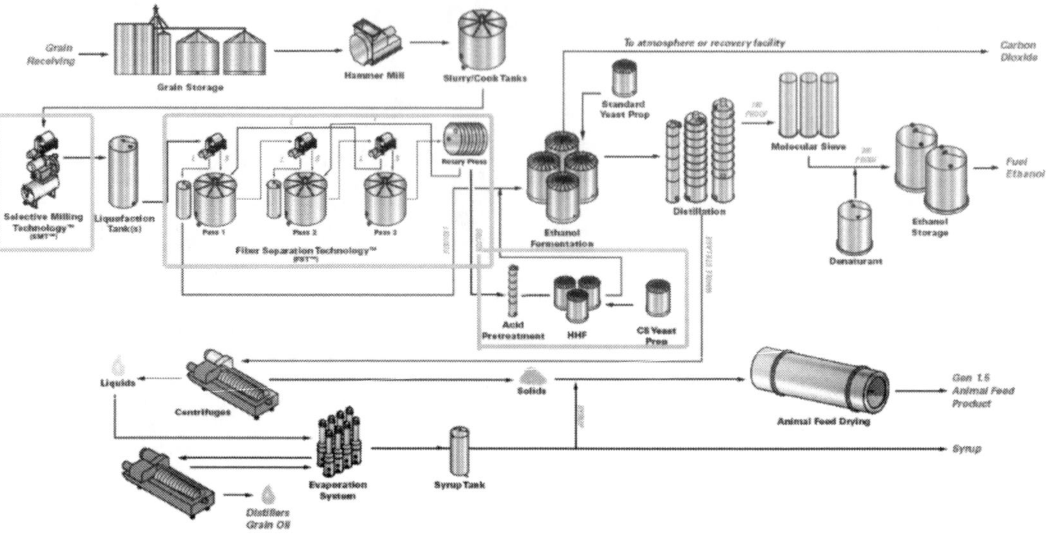

ICM the energy of innovation™

Next Generation Ethanol & Co-Product

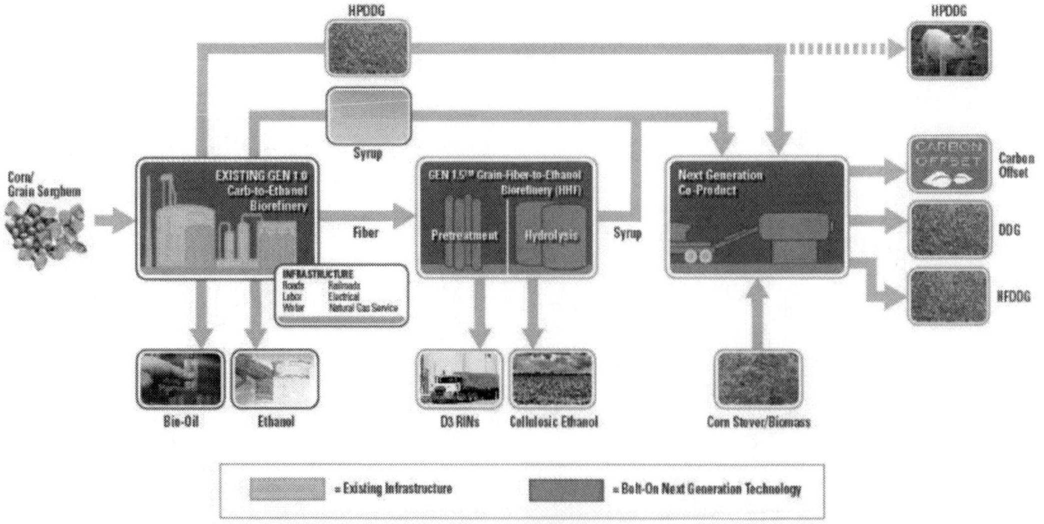

ICM the energy of innovation™

ICM, Inc. Proprietary/Confidential Commercial and Trade Secret Information protected by Contractual Secrecy Agreement. ICM Trade Secret Information protected by U.S. Economic Espionage Act and Uniform Trade Secrets Act. Patents Pending for Selective Milling Technology™, Fiber Separation Technology™ and Generation 1.5 Grain Fiber to Cellulosic Ethanol Technology™. Copying and Distribution Prohibited. © 2016 ICM, Inc. All Rights Reserved.

Non-Contact Water Use

the energy of innovation™

Chemical Use is Important
Gen 1.5 Ethanol with SMT™ and FST™

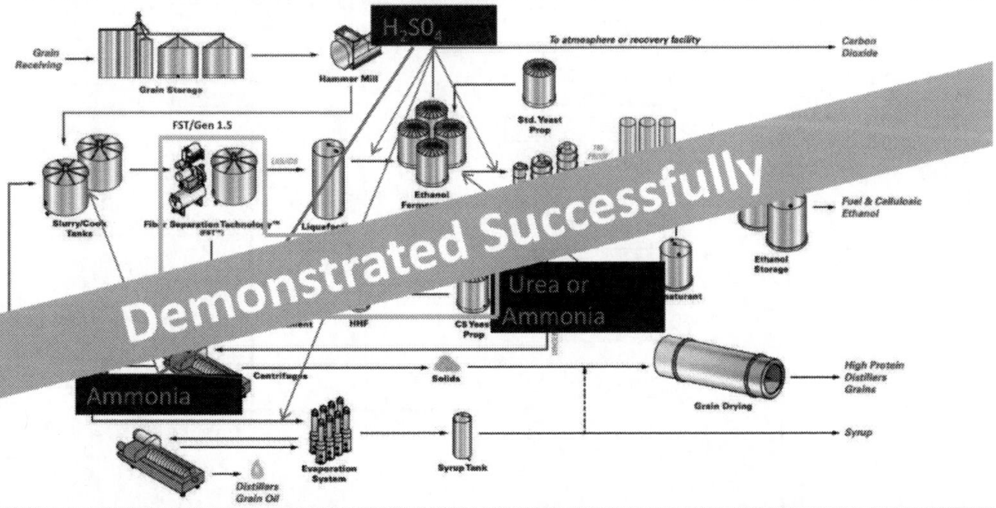

the energy of innovation™

ICM, Inc. Proprietary/Confidential Commercial and Trade Secret Information protected by Contractual Secrecy Agreement. ICM Trade Secret Information protected by U.S. Economic Espionage Act and Uniform Trade Secrets Act. Patents Pending for Selective Milling Technology™, Fiber Separation Technology™ and Generation 1.5 Grain Fiber to Cellulosic Ethanol Technology™. Copying and Distribution Prohibited. © 2016 ICM, Inc. All Rights Reserved.

Co-Located Model for Ethanol Production

Patent Pending

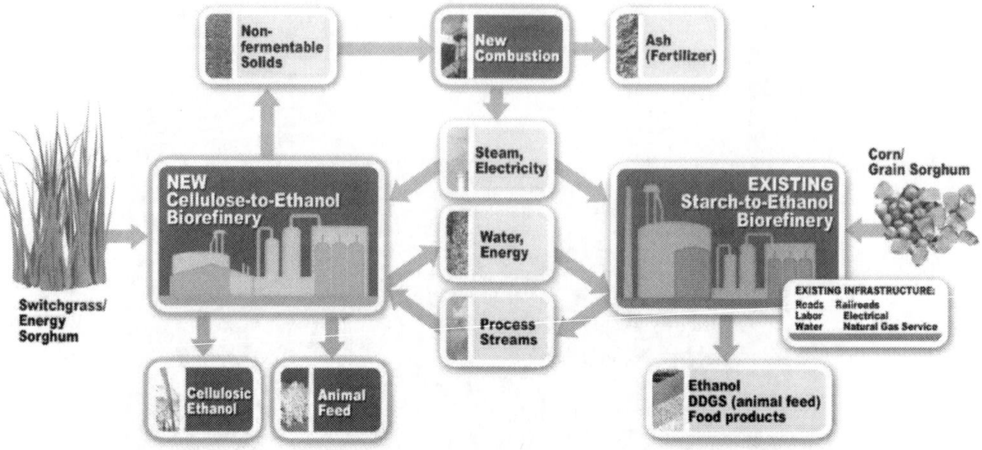

ICM the energy of innovation™

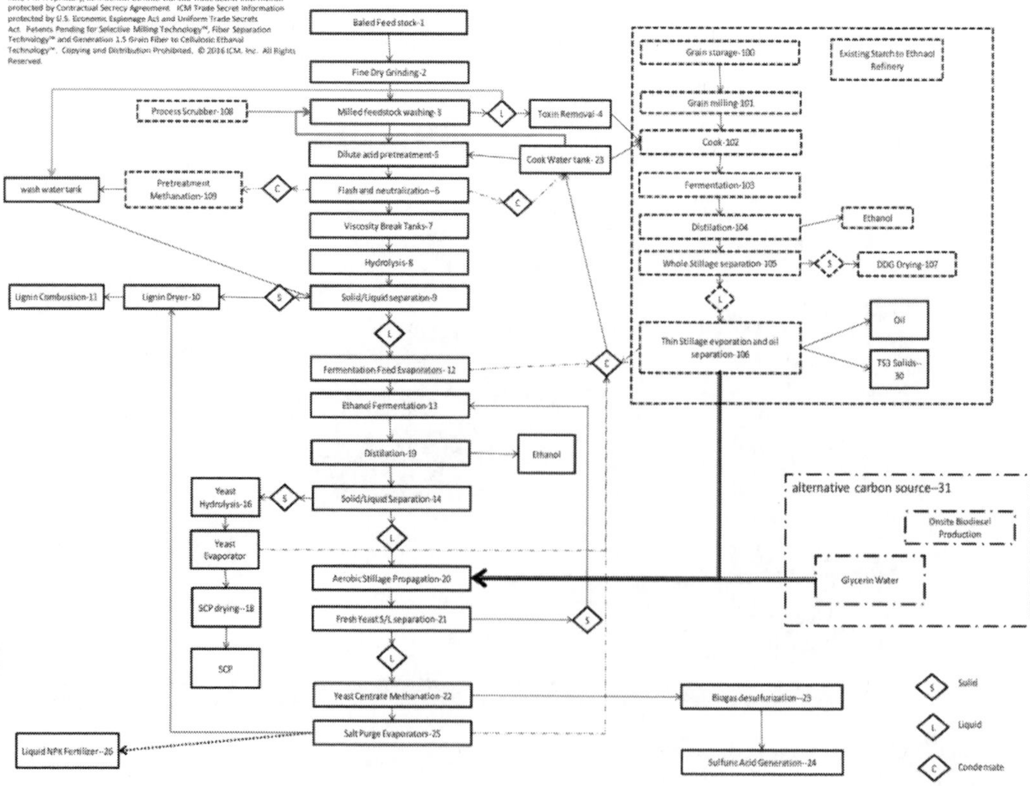

Generation 2.0: Biomass to Ethanol and Co-Products
Patent Pending

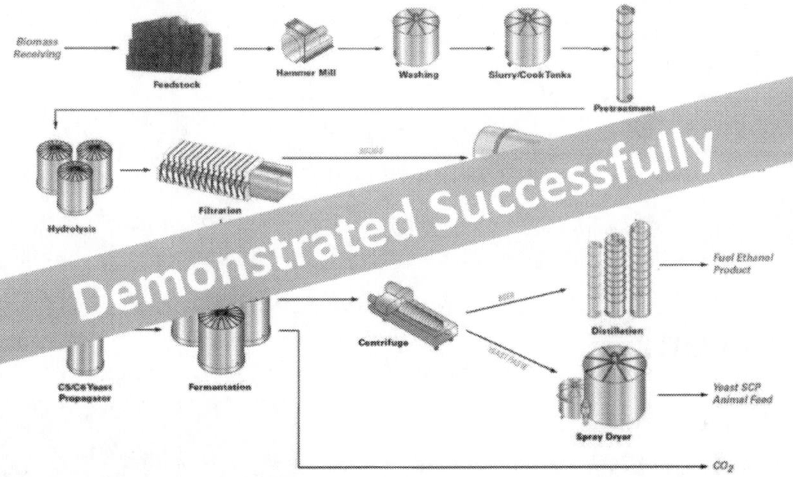

ICM the energy of innovation™

Summary

- Corn/Milo Ethanol Production More Efficient
- Water Consumption Down
  - Greatest Opportunity Non-Contact Water
- Energy Consumption Up
  - Biorefinery with More Unit Operations
- Energy/Gallon EtOH Down
  - More Efficient Design/Equipment
- Can Biomass Conversion Achieve Its Promise

ICM the energy of innovation™

Thank You!

Douglas B. Rivers, Ph.D.
2811 S. 11th Street, Ste 100
St. Joseph, MO 64503
Phone: 316.977.8502

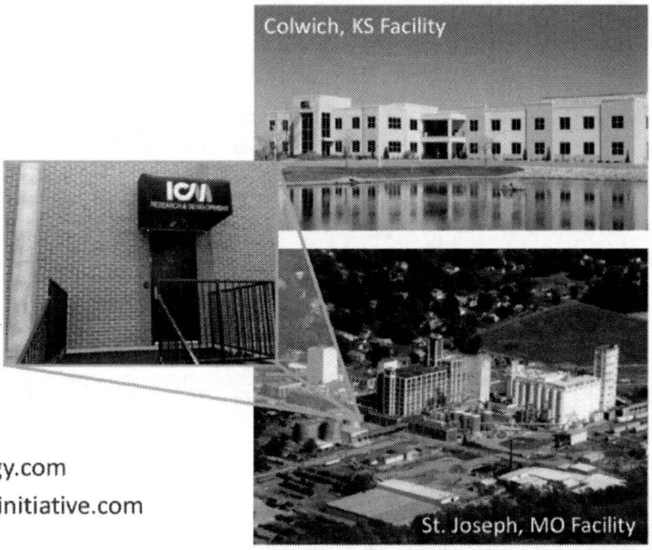

Helpful websites:

ICM – icminc.com
Growth Energy – growthenergy.com
Urban Air Initiative – urbanairinitiative.com
Ethanol – drivingethanol.org

ICM *the energy of innovation*™

Lessons Learned
Development and Scale-up in Aqueous Systems

James J. Grant III, P.E.

June 8, 2016

2016 Process Development Symposium | Providence, RI | June 7-9, 2016

DUPONT®

Agenda

- Process Development & Scale-up
- Aerobic Fermentation
- Evaporation of Water

2016 Process Development Symposium | Providence, RI | June 7-9, 2016

Process Development Timeline

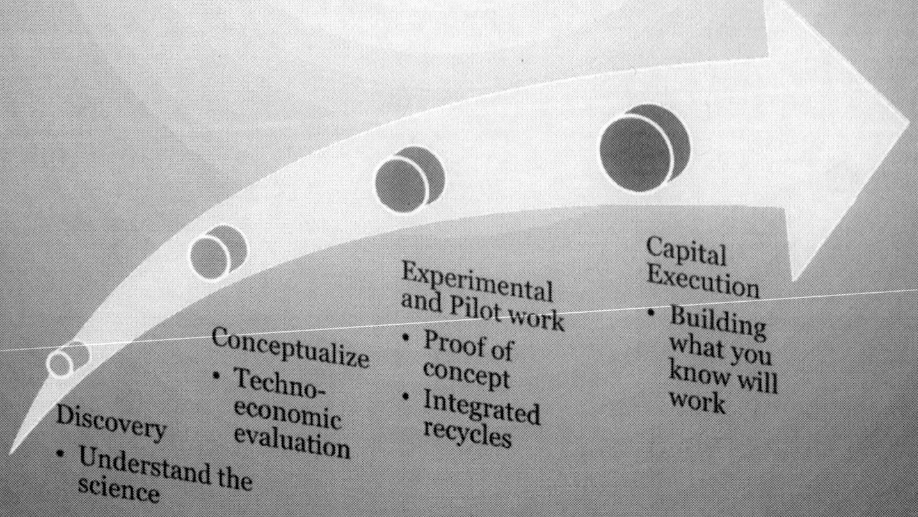

Process Development & Scale-up Questions to Ask Yourself

- Excessive experimentation before conceptualizing?
- Have we fixed concept based on only lab work?
- Have we kept the scientist/researcher apart from the engineer?
- It's a neat idea but is it scalable?
- What about the water balance?
- We're recycling all this water – undesirables concentrating?
- Is it economical / can we sell product profitably ?

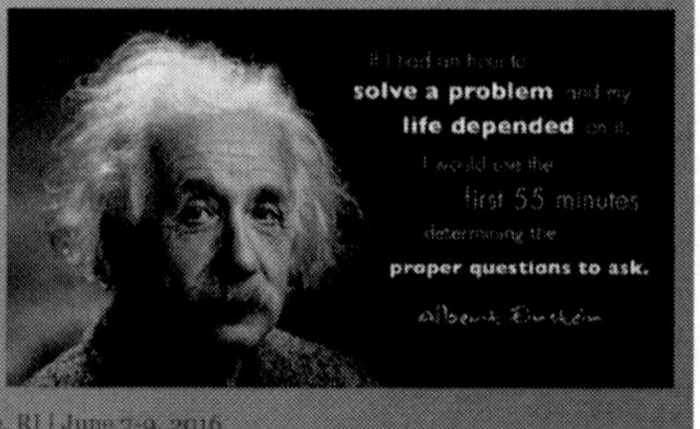

A Burning Question

What knowledge will significantly reduce the uncertainty of commercial success ?

Aerobic Fermentation Scale-up Issues

- Surface to Volume
- Oxygen Uptake Rates (OUR)
- Scale dependent design: Heat transfer / Surface:Vol
- Small scale : simple jacket may work
- Med-large scale : may need additional area / refrig
- Economics impacted by jacket design, cooling duty (OUR)

Aerobic Fermentation Scale-up Lessons

- Start by scaling down
- Impact of evaporative cooling
- Lab scale : high surface to volume
- Lab scale : high power input per unit volume
- Localized concentration effects - mixing
- Full scale fabrication issues – consider early on
    - wall thickness, jacket design, additional surface, refrigeration
- No. & size of fermenters – capital optimization
- No. & size also impact cooling cost/Opex

2016 Process Development Symposium | Providence, RI | June 7-9, 2016

Water, Water Everywhere! Yet not a drop to drink!

- Typical concentrations in the low weight %
- Final concentrations ~ 50-60+ %
- Tremendous energy requirements
- Water balance –be neither consumer nor producer
- Recycle water – may need to be treated
- Evaporation equipment – multi-effect ? MVR ?
- Think about these before first round of pilot work

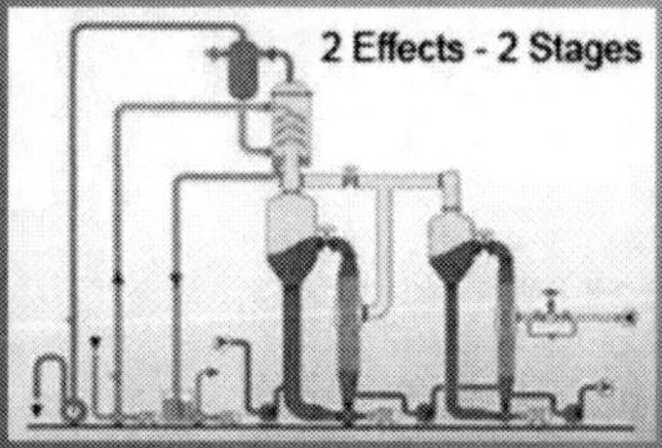

2016 Process Development Symposium | Providence, RI | June 7-9, 2016

Single Effect

Maximum Energy Usage

- 7% DSS feed concentration → 54% DSS product
- 50 Tonne/hr dilute feed
- 92 MM Btu/hr – 44 Tonne/hr vapor generation
- 44 Tonne/hr steam requirement – 1 Kg steam/Kg vapor
- 3.6 – 5.8 $MM /yr steam cost at $11 - $18 per tonne

Double Effect

10 psig → 0 psig

- 46 MM Btu/hr – 44 Tonne/hr vapor generation
- 22 Tonne/hr steam requirement
- 2 Kg vapor per Kg of steam usage
- Additional capital – one more evaporator @ 10°C ΔT
- Savings of 1.8 – 2.9 $MM / yr

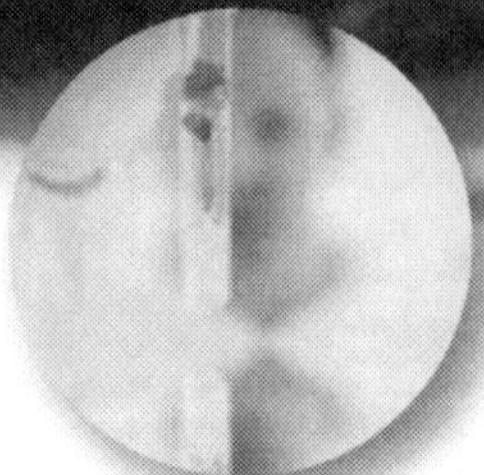

Triple Effect

10 psig → 0 psig → 8 psia

- 30 MM Btu/hr – 44 Tonne/hr vapor generation
- 14 Tonne/hr steam requirement
- 3 Kg vapor per Kg of steam usage
- Additional capital – two more evaporators @ 10°C ΔT
- Savings of 2.4 – 4 $MM / yr

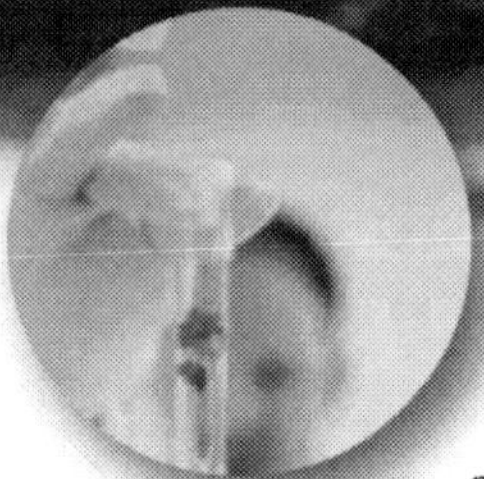

4 Effects

10 psig → 0 psig → 8 psia → 4psia

- 21 MM Btu/hr – 44 Tonne/hr vapor generation
- 10 Tonne/hr steam requirement
- 4.5 Kg vapor per Kg of steam usage
- Additional capital – 3 more evaporators @ 10°C ΔT
- Savings of 2.8 – 4.5 $MM / yr

MVR
Compression Ratio at 1.5

- 1.1 MW Electrical for 44 Tonne/hr vapor generation
- 0.4 – 0.8 $MM/yr electrical cost at $.05-.10 / kWh
- Additional Capital – fan/compressor
- Savings of 3.2 – 5 $MM / yr

Evaporation Equipment
Decisions, Decisions

- Look before you leap – Look before you pilot
- Temperature sensitivity might eliminate multi-effect
- Electricity price might eliminate MVR
- Limited scale might not justify MVR
- Boiler capacity may drive you to MVR
- Final effect temperature ok for water cooling?
- Site dependent decisions
- Development scale should consider these items
- Wetting rates and skin temperatures in pilot/full scale

Unsuccessful people make decisions based on their current situations. Successful people make decisions based on where they want to be.
Quoteistan.com

Lessons Learned

Development and Scale-up in Aqueous Systems

James J. Grant III, P.E.

June 8, 2016

2016 Process Development Symposium | Providence, RI | June 7-9, 2016

DUPONT®

Water, Water, Everywhere

Process and Economic Challenges of High Water Use Processes in Emerging Biological, Renewable and Environmental Technologies

John Ellersick, VP Projects

Oasys Water

PDS Process Development Symposium

OASYS WATER

Water, Water, Everywhere
General themes

- Introduction
- Reference processes
- Follow the money
- Making it all add up
- Presence of solids
- A Product, Waste, What else?
- The Order of Things
- Human Side of the Equation
- Wrapping it up

PDS Process Development Symposium

OASYS WATER

Introduction: How did we get here

- Chemical Engineers worked in conventional chemical industry since days of Arthur D Little
- All phases – gas / liquid / solids / hydrocarbon / aqueous, heterogeneous / homogenous
- Catalysts
- Some industries were always water based / intensive: mining, textiles, pulp & paper etc.
- Recent areas of industry that have grown have major aqueous focus: environmental, biological, renewables, "CleanTech"

Sustainability, BioTech Revolution

Introduction: Common Characteristics of Aqueous Processes

- Universal solvent
- Desired product(s) processed and/or produced in a liquid phase solution
- Suspensions, emulsions, heterogeneous
- Broad pH range
- Water % variability
- Presence of solids
- Separations required - difficult / complex / energy intensive
- Production of waste streams
- Byproducts / Coproducts
- Phase changes

Great for ChemE's!

Reference Processes
Oasys Water - Forward Osmosis

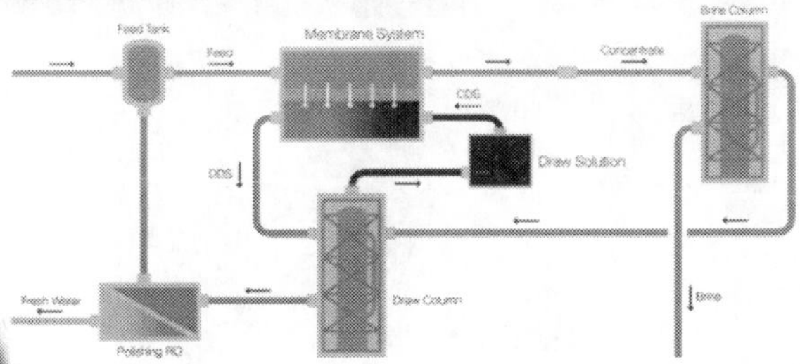

Reference Processes
Myriant Corp. – Bio-based Succinic Acid

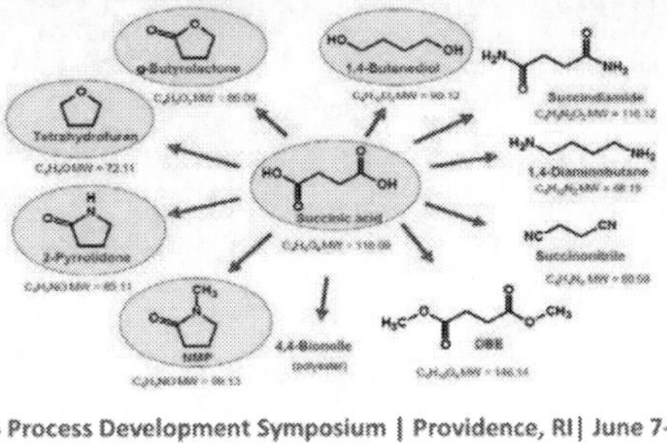

Reference Processes
Mascoma – Cellulosic Ethanol

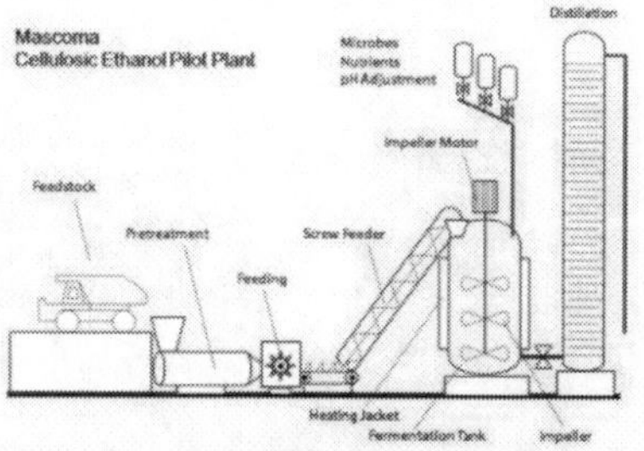

2016 Process Development Symposium | Providence, RI| June 7-9, 2016

Reference Processes
BioMetics - Biofine Process

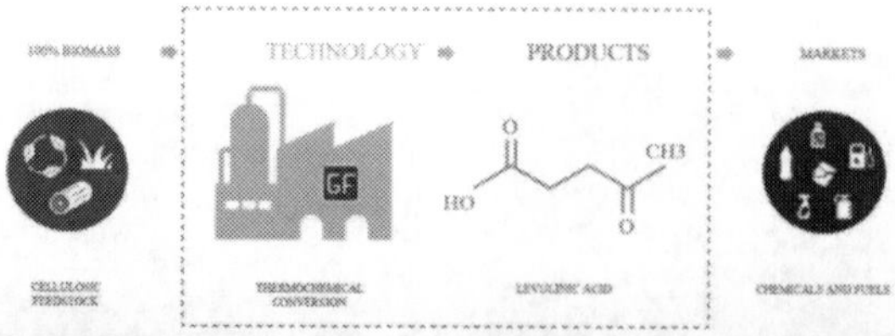

2016 Process Development Symposium | Providence, RI| June 7-9, 2016

Follow the Money: Without Process Economics You Don't Have a Process

- Economic models & evaluations need to encompass whole process to real battery limits (e.g. WWTP, disposal, water supply & treatment)
- Economics generally drive processes to more difficult modes of operation
- More water tends to drive processes to worse economics ... capital & operating costs
- Process optimization to achieve overall global economic improvements (not local)
  - What 3 ways can a process improvement be used?

Being too nice to your process will not be nice to your economics

2016 Process Development Symposium | Providence, RI| June 7-9, 2016

Follow the Money: Global Optimization & Process Impacts

- Membranes in abused water treatment
  - Membranes vs Capex vs Opex
- Fermentation in bio-industrial chemicals
  - Yield vs titer vs productivity
- Higher slurry concentrations in consolidated bioprocessing (CBP) cellulosic ethanol ...
  - Capex / OpEx vs operability
- Higher solids in cellulosic plastics
  - Capex / OpEx vs operability

lb water to lb product

2016 Process Development Symposium | Providence, RI| June 7-9, 2016

Making It All Add Up

- Best practice insists upon closing material balance at each scale
- Integrated pilot with extended runs = gold standard (and )
- At every scale, you can (& will) lose materials in different ways
- +Process units = +yield & product losses
  - Care about every waste stream and every loss
- Understand & design for all modes of operation

PDS Process Development Symposium

Just say no to single point design

2016 Process Development Symposium | Providence, RI| June 7-9, 2016

OASYS WATER

Making It All Add Up: Examples

- Heel & fermentation cleanout losses
- Every loss shows up somewhere else as a waste or an emission ... wwtp as plant bottleneck
- Steady state performance after lining out requires demonstration
- Technologies with tremendous organic loads not initially identified (large anaerobic biotreaters)
- Modeling balances during all different modes of operation – startup, steady state, batches, upset, shutdown, etc.

PDS Process Development Symposium

Diversity & Unsteady State Modeling

2016 Process Development Symposium | Providence, RI| June 7-9, 2016

OASYS WATER

Presence of Solids

- Generally overall process economics favor operation at higher solids
- Top 5 items making process scaleup difficult
- Velocities, settling, erosion, plugging, fouling
- Viscosity & non-Newtonian behavior
- Solids management & removal learnings expected to occur - major purpose of scaleup / demo plants
- Solids / moisture content behavior differs greatly between compounds

Design for highest solids, allows ops at lower levels if needed

2016 Process Development Symposium | Providence, RI| June 7-9, 2016

A Product, Waste, What Else?

- At what % is a product not worth recovering?
- At what cost must it be treated
- Alternative approaches
- Best solutions for waste situations often found by heading upstream
- Can you sell it, give it away?
- Segregation – strength, flow, other
- Reuse, recovery
- ZLD

2016 Process Development Symposium | Providence, RI| June 7-9, 2016

A Product, Waste, What Else?
Example Segregate & Treatment Options

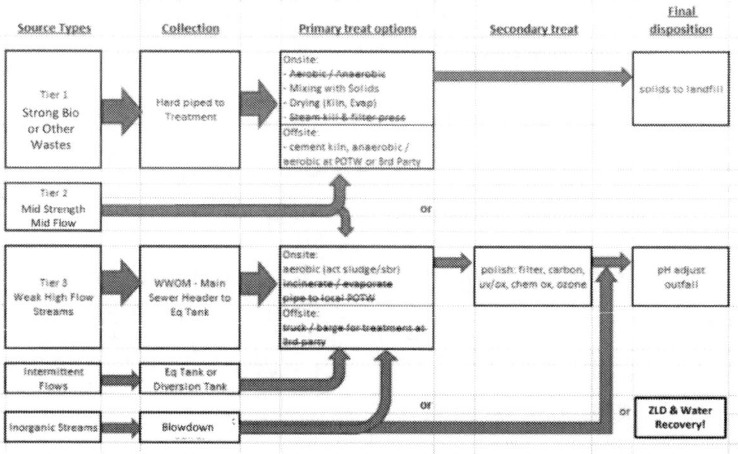

Approach	Design Basis Impact	Cap Cost Delta	Op Cost Delta	Disposition

The Order of Things: General Scale-up & Commercialization Takeaways

- Make the product, recover the product, sell the product, complete the whole process picture
- Scaling up, down ... process, economics
- Integration by steps (listen to Aspen/KISS)
- If you are not creating lessons learned, you might not be trying hard enough (fail fast)
- Bring a map
- Be (intellectually) honest & realistic about costs ...

Design for most demanding conditions, allow operations at lower levels if needed

2016 Process Development Symposium | Providence, RI| June 7-9, 2016

Lessons Learned ... on Lessons Learned

Best practices

- Make it a regular practice
- Positive, honest, forward looking
- Adjust processes, then KO next projects with a look back
- Define solutions, generate action plan, track progress
- Then move on

Why do we keep generating more lessons learned?

- Inherent in Technology development
- Perfect is the enemy of the good
- Fail fast approach
- Continuous improvement

Best practices & learnings are IP

2016 Process Development Symposium | Providence, RI| June 7-9, 2016

Waterfall .. You Need a Road Map

COST IMPROVEMENTS & CHANGES WITH TECHNOLOGY DEVELOPMENT

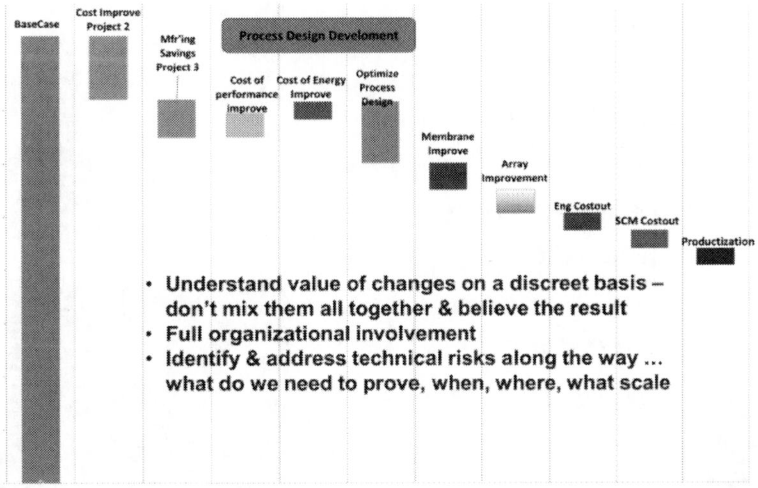

- **Understand value of changes on a discreet basis – don't mix them all together & believe the result**
- **Full organizational involvement**
- **Identify & address technical risks along the way ... what do we need to prove, when, where, what scale**

2016 Process Development Symposium | Providence, RI| June 7-9, 2016

Cost Estimation – Caution with Scaling up

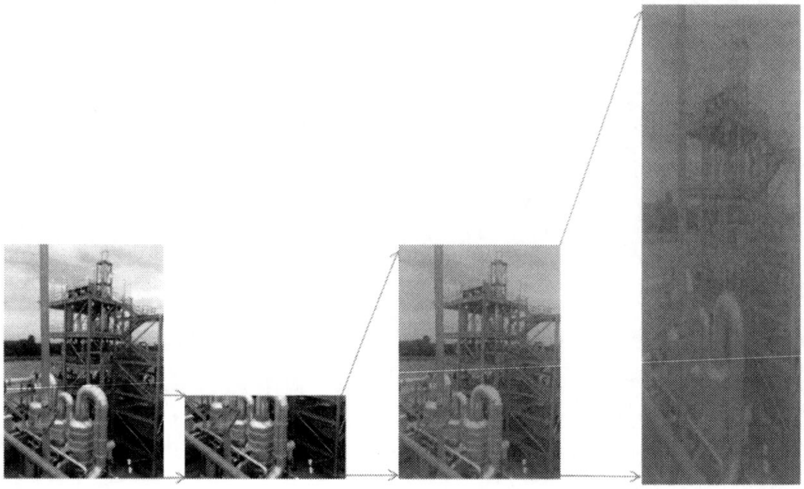

Human Side of the Equation: Scaleup Also Means Scaling Organizations

- Transitioning from R&D / pilot scale to commercial scale systems & fabrication
- Supply chain development & scale-up
- Expanding sales organization
- Growing engineering & technical team…
- Partners

Human Side of the Equation: Partners

PDS Process Development Symposium

2016 Process Development Symposium | Providence, RI| June 7-9, 2016

OASYS WATER

Human Side of the Equation Mission, Values, Motivation

Oasys Values

- Think Forward
- Be Excellent
- Execute Safely
- Create Value
- Win Together

PDS Process Development Symposium

2016 Process Development Symposium | Providence, RI| June 7-9, 2016

OASYS WATER

Wrapping it up

- Aqueous processes will continue to grow in importance & as a proportion of ChemE focus
- Key challenges include:
  - Driving process to more difficult conditions to achieve best economics,
  - Global optimization
  - Solids
  - Weak products / strong wastes
  - Tightening up and managing the water balance
- Attend to the human side of the equation

PDS Process Development Symposium

OASYS WATER

Thanks & Questions

PDS Process Development Symposium

OASYS WATER

Liquid-Liquid Extraction for Recovery of Chemicals from Dilute Aqueous Streams

Donald Glatz

Koch Modular Process System, LLC

PDS Process Development Symposium

K M P S Extraction Technology Group www.modularprocess.com

2016 Process Development Symposium | Providence, RI| June 7-9, 2016

Countercurrent Extraction Column

KMPS Koch Modular Process Systems, LLC.

MODULAR PROCESS SYSTEMS | DISTILLATION | LIQUID EXTRACTION | ENGINEERING | PILOT TESTING

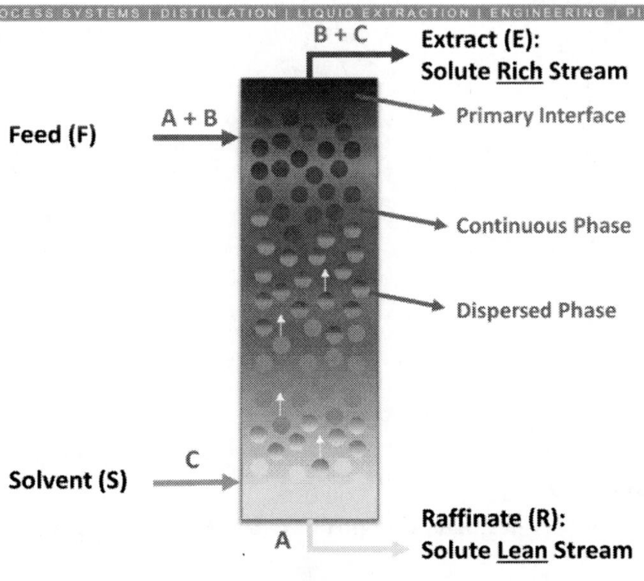

PDS Process Development Symposium

www.modularprocess.com
Telephone: (201) 267-8670

2

2016 Process Development Symposium | Providence, RI| June 7-9, 2016

LLE Equilibrium and Operating Lines

KMPS Koch Modular Process Systems, LLC.

MODULAR PROCESS SYSTEMS | DISTILLATION | LIQUID EXTRACTION | ENGINEERING | PILOT TESTING

Solute Free Basis

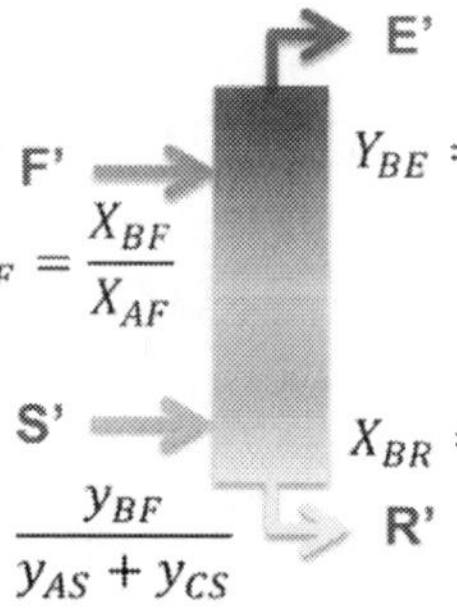

$$Y_{BE} = \frac{y_{BE}}{y_{AR} + y_{CE}}$$

$$X_{BF} = \frac{X_{BF}}{X_{AF}}$$

$$X_{BR} = \frac{x_{BR}}{x_{AR} + x_{CR}}$$

$$Y_{BS} = \frac{y_{BF}}{y_{AS} + y_{CS}}$$

$$F' = F(x_{AR})$$
$$S' = S(y_{AS}y_{CS})$$
$$E' = E(y_{AE}y_{CE})$$
$$R' = R(x_{AR}x_{CR})$$

Graphical Solution

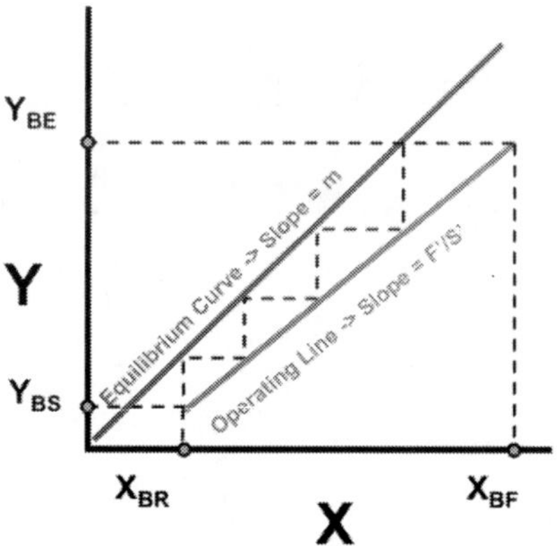

$$m = \frac{Y_{B^*}}{X_{B^*}} \left(\begin{array}{c}\text{Distribution Coefficient}\\ \text{on Solute Free Basis}\end{array}\right)$$

PDS Process Development Symposium

Typical LLE Equilibrium Curve

KMPS Koch Modular Process Systems, LLC.

MODULAR PROCESS SYSTEMS | DISTILLATION | LIQUID EXTRACTION | ENGINEERING | PILOT TESTING

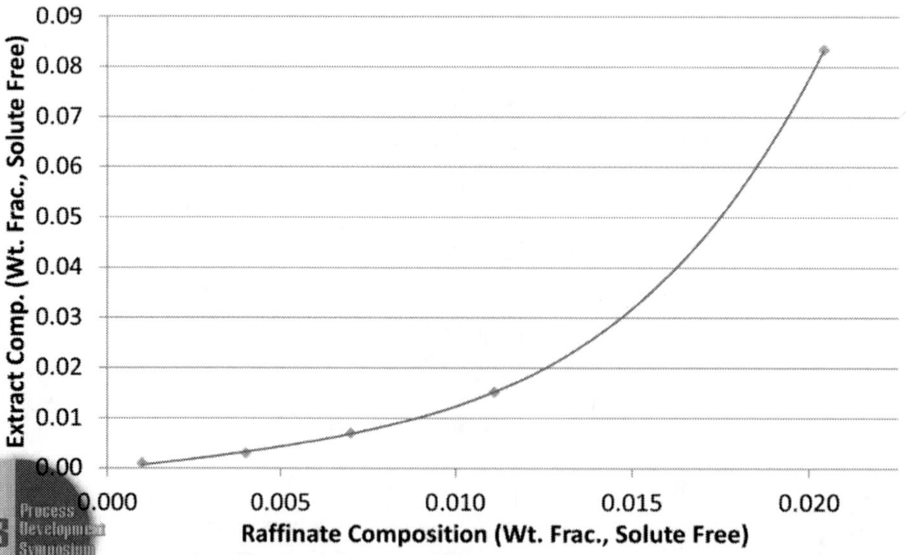

PDS Process Development Symposium

Solvent Selection

KMPS Koch Modular Process Systems, LLC
MODULAR PROCESS SYSTEMS | DISTILLATION | LIQUID EXTRACTION | ENGINEERING | PILOT TESTING

- Literature – DECHEMA, Perry's Chemical Engineers Handbook
- Internet search
- Understanding the chemistry of the process
- Experience – consulting with experts
- Laboratory Experiments:
  1. Quickly screen solvents
  2. Generate equilibrium data
  3. Observe phase behavior

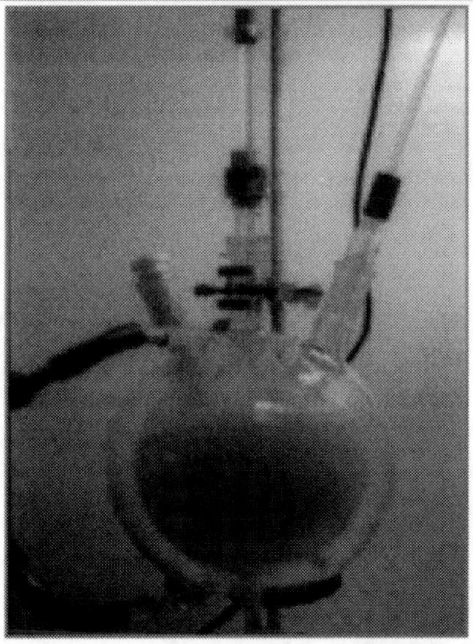

PDS Process Development Symposium

www.modularprocess.com
Telephone: (201) 267-8670

Bench Scale Test Apparatus for Generating Liquid-Liquid Equilibrium Data

KMPS Koch Modular Process Systems, LLC
MODULAR PROCESS SYSTEMS | DISTILLATION | LIQUID EXTRACTION | ENGINEERING | PILOT TESTING

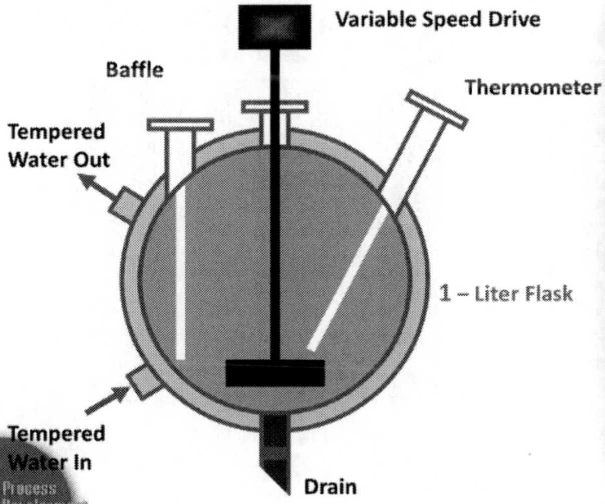

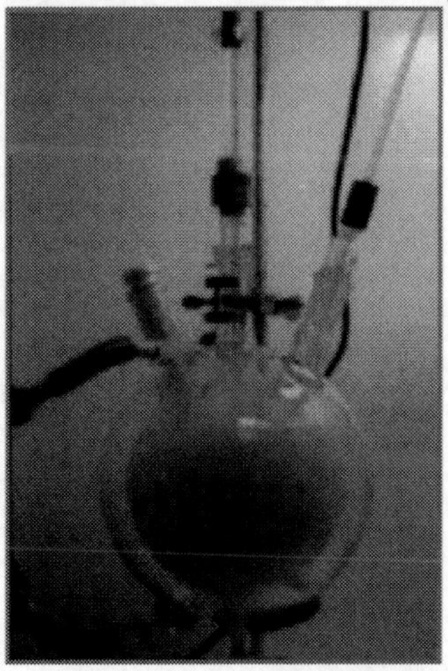

PDS Process Development Symposium

www.modularprocess.com
Telephone: (201) 267-8670

Kremser Equation

KMPS Koch Modular Process Systems, LLC.

MODULAR PROCESS SYSTEMS | DISTILLATION | LIQUID EXTRACTION | ENGINEERING | PILOT TESTING

$$n = \frac{\mathrm{LOG}\left[\left(\frac{x_f - \frac{y_s}{m}}{x_n - \frac{y_s}{m}}\right)\left(1 - \frac{1}{E}\right) + \frac{1}{E}\right]}{\mathrm{LOG}\,E}$$

Where:
- n = Number of theoretical stages required
- x_f = Conc. of solute in feed on solute free basis
- x_n = Conc. of solute in raffinate on solute free basis
- y_s = Conc. of solute in solvent on solute free basis
- m = Distribution coefficient
- E = Extraction factor = (m)(S/F)

PDS Process Development Symposium

www.modularprocess.com
Telephone: (201) 267-8670

2016 Process Development Symposium | Providence, RI | June 7-9, 2016

7

Recovery of Acetic Acid from Water Using a Low Boiling Solvent

KMPS Koch Modular Process Systems, LLC.

MODULAR PROCESS SYSTEMS | DISTILLATION | LIQUID EXTRACTION | ENGINEERING | PILOT TESTING

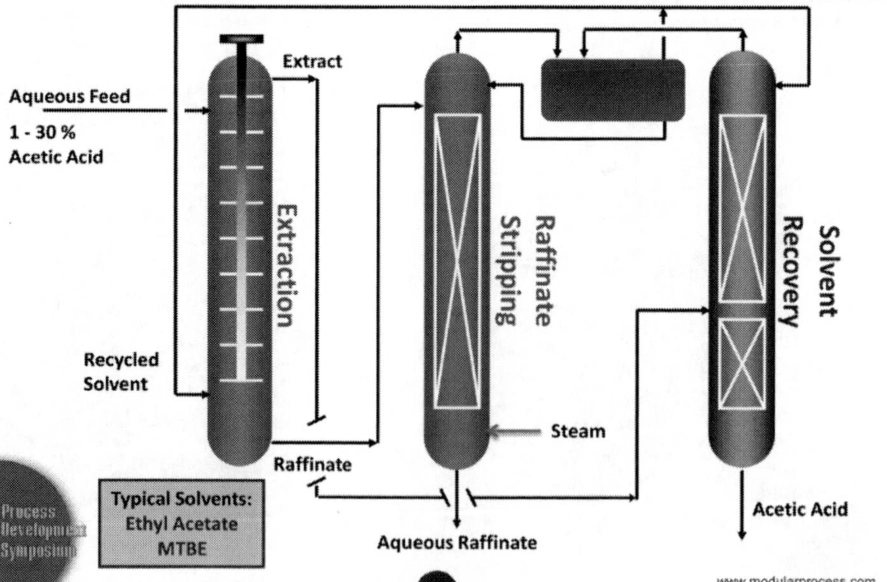

PDS Process Development Symposium

www.modularprocess.com
Telephone: (201) 267-8670

2016 Process Development Symposium | Providence, RI | June 7-9, 2016

8

Determine Solvent to Feed Ratio

Kremser Calculation - Column Performance Evaluation

MTBE - Acetic Acid - Water

No.	Xf	Xn	Ys	S	F	m	E	A	Ns
98% Acetic Acid Extraction									
	0.050000	0.0010000	0.000000	2.50	1.00	0.80	2.00	50.0	4.67
	0.050000	0.0010000	0.000000	2.00	1.00	0.80	1.60	50.0	6.31
	0.050000	0.0010000	0.000000	1.60	1.00	0.80	1.28	50.0	9.97
	0.050000	0.0010000	0.000000	1.40	1.00	0.80	1.12	50.0	16.17
98% Acetic Acid Extraction - Some Acetic Acid in Solvent									
	0.050000	0.0010000	0.001000	1.60	1.00	0.80	1.28	-195.0	#NUM!
	0.050000	0.0010000	0.000500	1.60	1.00	0.80	1.28	131.7	13.72
	0.050000	0.0010000	0.000100	1.60	1.00	0.80	1.28	57.0	10.47

Xf = solute conc. in feed / (1 - solute conc. in feed)
Xn = solute conc. in raffinate / (1 - solute conc. in raffinate)
Ys = solute conc. in solvent / (1 - solute conc. in solvent)
S = mass solvent rate x (1 - solute conc. in solvent)
F = mass feed rate x (1 - solute conc. in feed)
m = distribution coefficient
E = extraction factor (S/F x m)
A = (Xf - (Ys/m)) / (Xn - (Ys/m))
Ns = number of theoretical stages via Kremser Equation

PDS Process Development Symposium

Determine Process Feasibility

KMPS Koch Modular Process Systems, LLC

MODULAR PROCESS SYSTEMS | DISTILLATION | LIQUID EXTRACTION | ENGINEERING | PILOT TESTING

- Calculate number theoretic stages vs solvent to feed ratio
- Estimate material balance around LLE column
- Provide feed rates/concentrations for distillation and stripping columns
- Optimize distillation and steam stripper via computer simulations
- Size equipment and prepare capital cost estimate
- From utilities generated by simulations, estimate the operating cost

PDS Process Development Symposium

www.modularprocess.com
Telephone: (201) 267-8670

Liquid-Liquid Extraction Scale-Up

KMPS Koch Modular Process Systems, LLC.

MODULAR PROCESS SYSTEMS | DISTILLATION | LIQUID EXTRACTION | ENGINEERING | PILOT TESTING

Best method of design:
Pilot testing followed by empirical scale-up

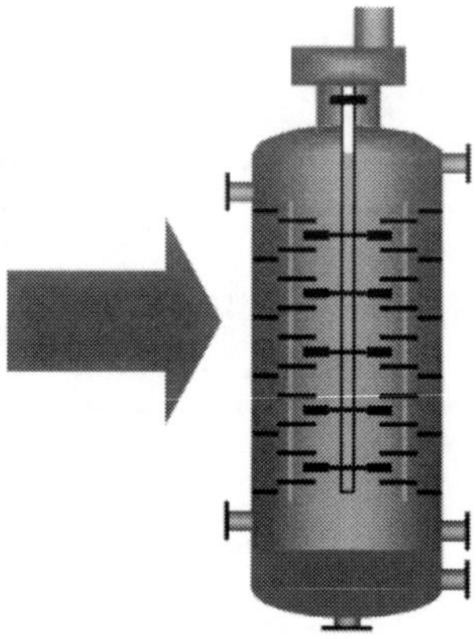

PDS Process Development Symposium

Pilot Testing – Step by Step Optimization

KMPS Koch Modular Process Systems, LLC.

MODULAR PROCESS SYSTEMS | DISTILLATION | LIQUID EXTRACTION | ENGINEERING | PILOT TESTING

- Select column type and set-up column (multiple inlets)
- Select continuous and dispersed phases
- Select initial operating conditions (capacity – S/F – No. Stages)
- Operate at increasing agitation speed (flooding)
- Based on results, vary:
  1. Continuous phase
  2. S/F vs. No. Stages
  3. Capacity
  4. Other Variables (temp, pH, etc.)

PDS Process Development Symposium

Application - Recovery of Carboxylic Acids from Fermentation Broth

KMPS Koch Modular Process Systems, LLC.

MODULAR PROCESS SYSTEMS | DISTILLATION | LIQUID EXTRACTION | ENGINEERING | PILOT TESTING

Broth generated from cellulosic materials – approx. 5% acids

LLE Goal: To achieve >95% recovery (high purity) and minimize solvent usage

Ethyl acetate selected solvent – but emulsified easily

Preliminary Data in RDC Columns

Difficult operation due to emulsification

< 90% acid recovery

High S/F ratio – 2.0

KARR® Column is preferred choice because
This design handles systems that emulsify

PDS Process Development Symposium

Pilot KARR® Column for Carboxylic Acids 1" diameter x 12' Plate Stack

KMPS Koch Modular Process Systems, LLC.

MODULAR PROCESS SYSTEMS | DISTILLATION | LIQUID EXTRACTION | ENGINEERING | PILOT TESTING

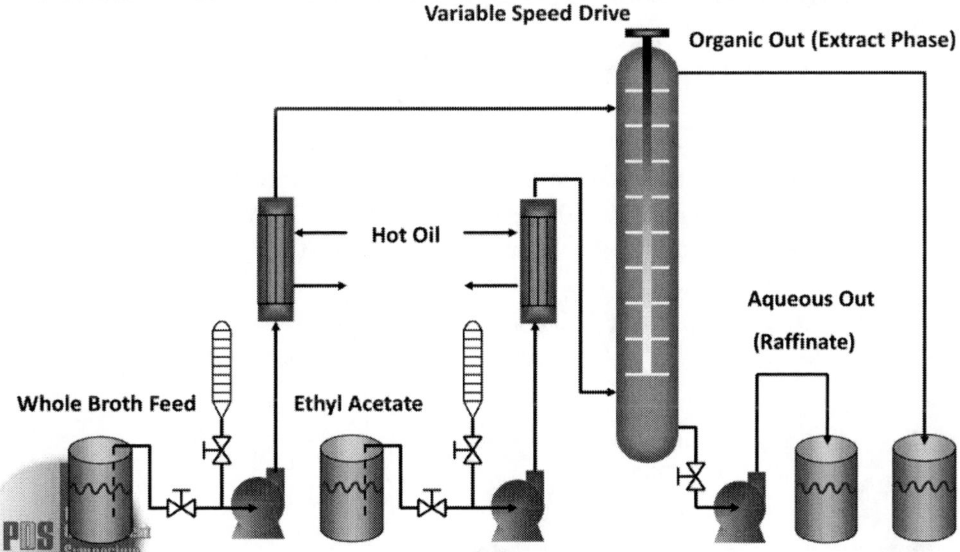

PDS Symposium

Results from Extraction Testing

KMPS Koch Modular Process Systems, LLC.

MODULAR PROCESS SYSTEMS | DISTILLATION | LIQUID EXTRACTION | ENGINEERING | PILOT TESTING

- KARR® Column is the correct design for this process which failed in RDC column.
- S/F ratio improved to 1.5 (client expected 2)
- 98-99% recovery at 40-45 C
- Design capacity 650 GPH/ft^2
- Complete system (LLE + distillations) designed, fabricated, and installed
- LLE testing identified "unknown" component

PDS Process Development Symposium

KARR® Column Pilot Plant Scale-up Procedure: Carboxylic Acid Extraction

KMPS Koch Modular Process Systems, LLC.

MODULAR PROCESS SYSTEMS | DISTILLATION | LIQUID EXTRACTION | ENGINEERING | PILOT TESTING

- Select optimal run from test results
  - Run Number 11:
    - Solvent to Feed Ratio (S/F) = 1.5
    - Specific Throughput (Q) = 650 GPH/FT^2
- Production column design
  - Diameter – direct scale-up based on specific throughput
  - Broth Feed = 4974 lb/hr, Ethyl Acetate = 7461 lb/hr
  - Area = (Feed rate + Solvent Rate) / 650 GPH/FT^2
  - Height: $H_{COMM} = f\,(H)_{PILOT}$
  - Agitation Speed: $SPM_{COMM} = f\,(SPM)_{PILOT}$

PDS Process Development Symposium

KARR® Column Pilot Plant Scale-up Procedure
Carboxylic Acid Extraction from Broth

MODULAR PROCESS SYSTEMS | DISTILLATION | LIQUID EXTRACTION | ENGINEERING | PILOT TESTING

- $H_{COMM} = (D_{COMM} / D_{PILOT})^{0.38} \times H_{PILOT}$
- $H_{COMM} = (22/1)^{0.38} \times (12 \text{ feet}) = 38 \text{ feet}$
- $SPM_{COMM} = (D_{PILOT} / D_{COMM})^{0.14} \times SPM_{PILOT}$
- $SPM_{COMM} = (1/22)^{0.14} \times (50 \text{ SPM}) = 32 \text{ SPM}$
- Where:
    - H_{COMM} = Height Commercial Column
    - H_{PILOT} = Height Pilot Column
    - D_{COMM} = Diameter Commercial Column
    - D_{PILOT} = Diameter Pilot Column
    - SPM_{COMM} = Commercial Strokes Per Minute
    - SPM_{PILOT} = Pilot Strokes Per Minute

17

2016 Process Development Symposium | Providence, RI | June 7-9, 2016

www.modularprocess.com
KARR® and SCHEIBEL® are registered trademarks of Koch-Glitsch, LP.
Telephone: (201) 267-8670

KARR® Column Design
Carboxylic Acid Extraction from Broth

MODULAR PROCESS SYSTEMS | DISTILLATION | LIQUID EXTRACTION | ENGINEERING | PILOT TESTING

- **Diameter = 22" (D_1)**
- **Expanded Ends**
  Diameter = 44" (D_2)
- **Plate Stack = 38'-0" (A)**
- **Overall Height = 50'-6" (B)**

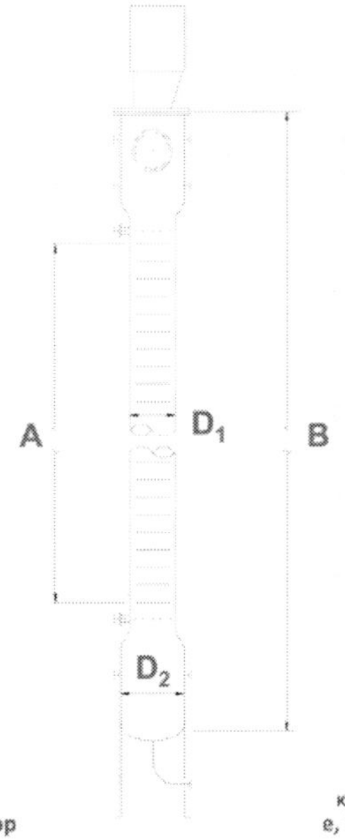

2016 Process Develop... e, RI | June 7-9, 2016

www.modularprocess.com
KARR® and SCHEIBEL® are registered trademarks of Koch-Glitsch, LP.
Telephone: (201) 267-8670

Batch Neutralization and Water Wash

KMPS Koch Modular Process Systems, LLC

MODULAR PROCESS SYSTEMS | DISTILLATION | LIQUID EXTRACTION | ENGINEERING | PILOT TESTING

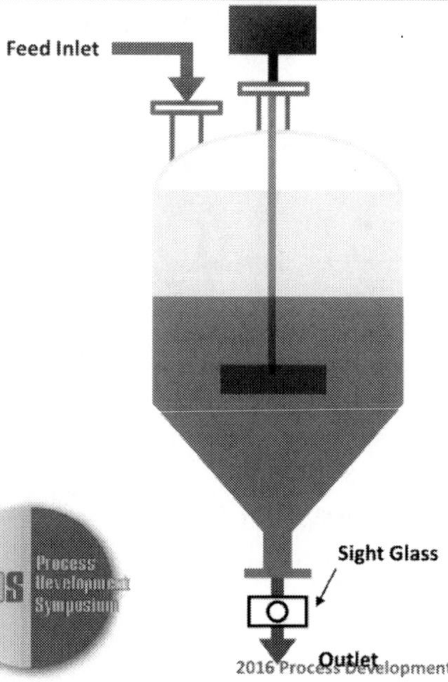

Characteristics

- Mix – Settle – Phase separate in a single tank
- Requires multiple water additions for more than one stage (crossflow operation)
- Typically removal of inorganic compounds from organic feed streams
- Sometime neutralization with acid or base followed by water washes to remove salts

Cross Flow Extraction

KMPS Koch Modular Process Systems, LLC

MODULAR PROCESS SYSTEMS | DISTILLATION | LIQUID EXTRACTION | ENGINEERING | PILOT TESTING

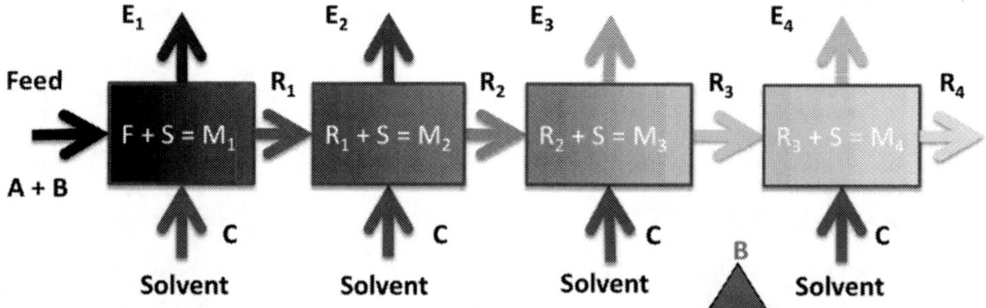

LEGEND	
A, B	Components in feed
C	Component in solvent
R_x	Raffinate from stage x
E_x	Extract (made up of components B & C)
M_x	Composition of two phase mixture

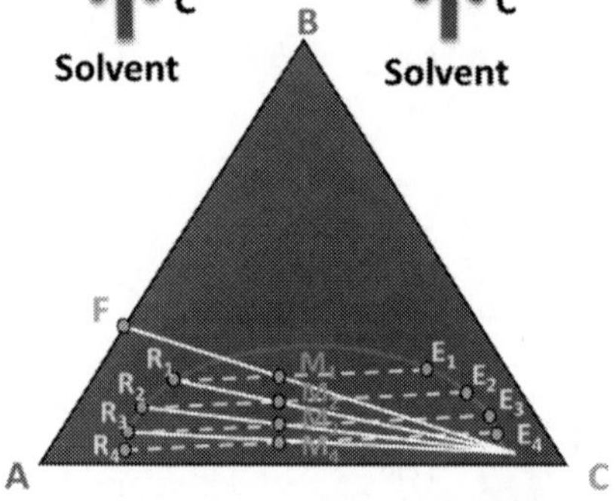

Mix / Decant Tank

KMPS Koch Modular Process Systems, LLC.

MODULAR PROCESS SYSTEMS | DISTILLATION | LIQUID EXTRACTION | ENGINEERING | PILOT TESTING

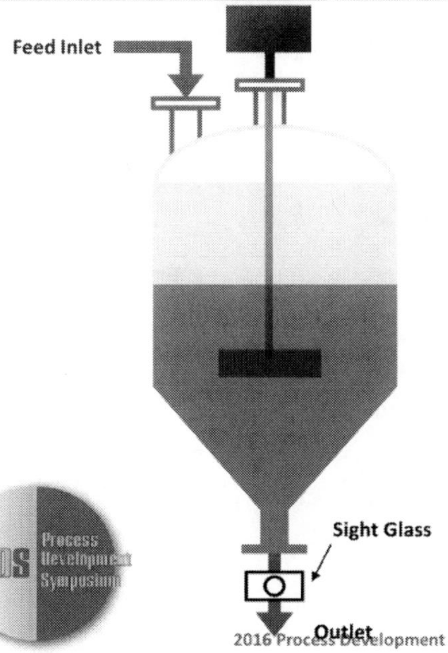

Typical Applications

- Water to remove K-salts from phenolic compounds
- Water to extract undesirable impurities from gasoline
- Water to extract sulfur compounds from agricultural chemical
- Water to extract monomer from polymer solution
- Water to extract metal ions from polymer solutions
- Water to extract MeOH and glycerin from methyl esters (biodiesel)
- Water to extract MeOH from acrylates

PDS Process Development Symposium

21

www.modularprocess.com
Telephone: (201) 267-8670

2016 Process Development Symposium | Providence, RI| June 7-9, 2016

Neutralization/Washing of Acid or Base or Polar Compounds from Organic Stream

KMPS Koch Modular Process Systems, LLC.

MODULAR PROCESS SYSTEMS | DISTILLATION | LIQUID EXTRACTION | ENGINEERING | PILOT TESTING

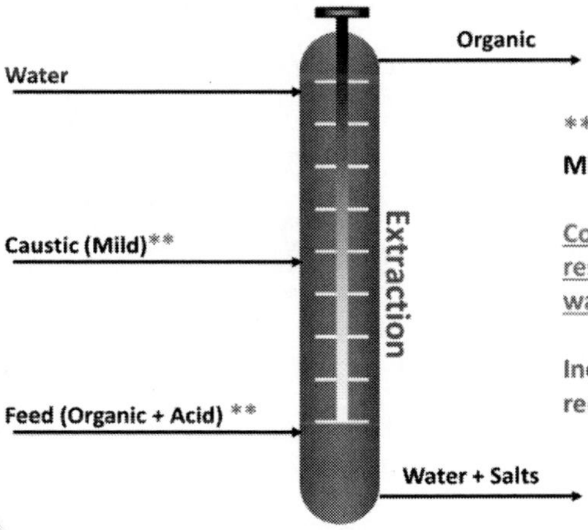

Organic Feed could be basic. Mid-Feed would be mild acid

Counter current extraction column results in significant reduction in water usage vs batch processing

Increasing the number of stages reduces the solvent to feed ratio

PDS Process Development Symposium

22

www.modularprocess.com
Telephone: (201) 267-8670

2016 Process Development Symposium | Providence, RI| June 7-9, 2016

Water Extraction (Washing)

KMPS Koch Modular Process Systems, LLC

MODULAR PROCESS SYSTEMS | DISTILLATION | LIQUID EXTRACTION | ENGINEERING | PILOT TESTING

Characteristics

- Typically distribution coefficient is very high (>100)
- Very low water : feed ratio are possible (0.03 : 1.0)
- However, due to low phase ratio, excellent phase contact is required
- High agitation and minimization of back-mixing must be achieved
- Columns such as the SCHEIBEL Column provide excellent mixing and baffling to minimize back mixing

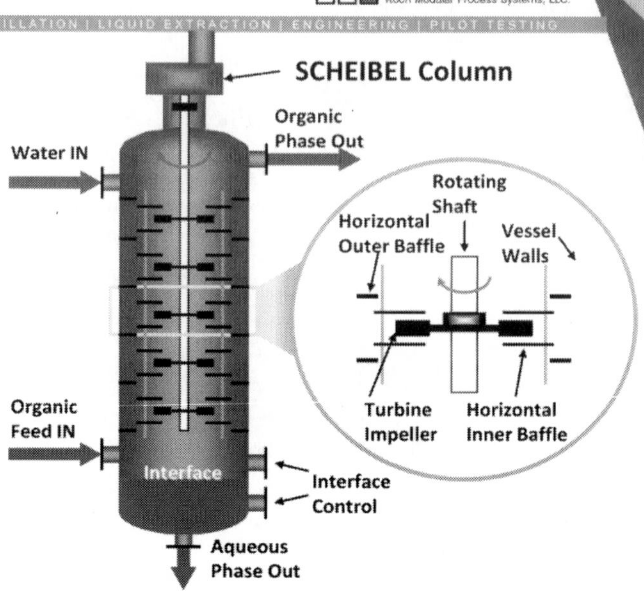

PDS Process Development Symposium

www.modularprocess.com
Telephone: (201) 267-8670

Biodiesel Extraction Process

- Shake tests in 2-liter round bottom flask
- S/F ratio = 0.05
- Feed concentration
  Methanol= 2.6%
  Glycerin= 6000 ppm
- Target raffinate concs.
  Methanol < 0.1%
  Glycerin < 200 ppm

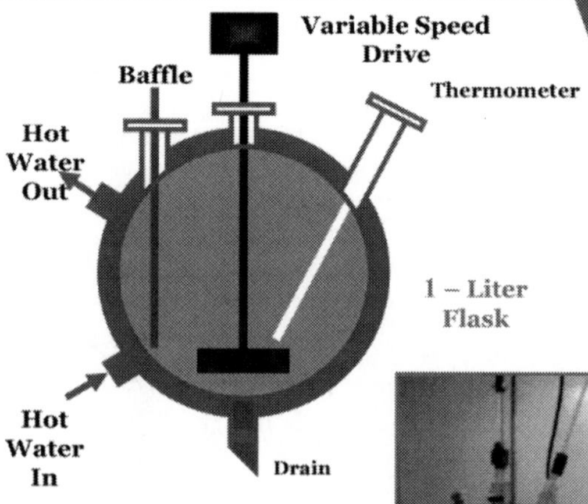

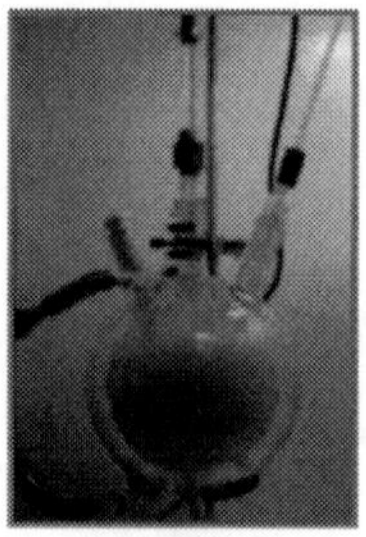

PDS Process Development Symposium

Results from shake test

- <200 ppm glycerin in two stages
- <0.1% methanol in four stages
- Slow phase separation
- Emulsion at interface

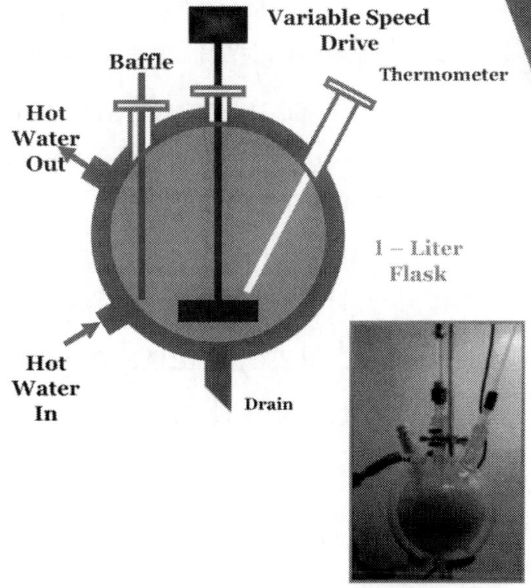

Pilot Plant Testing

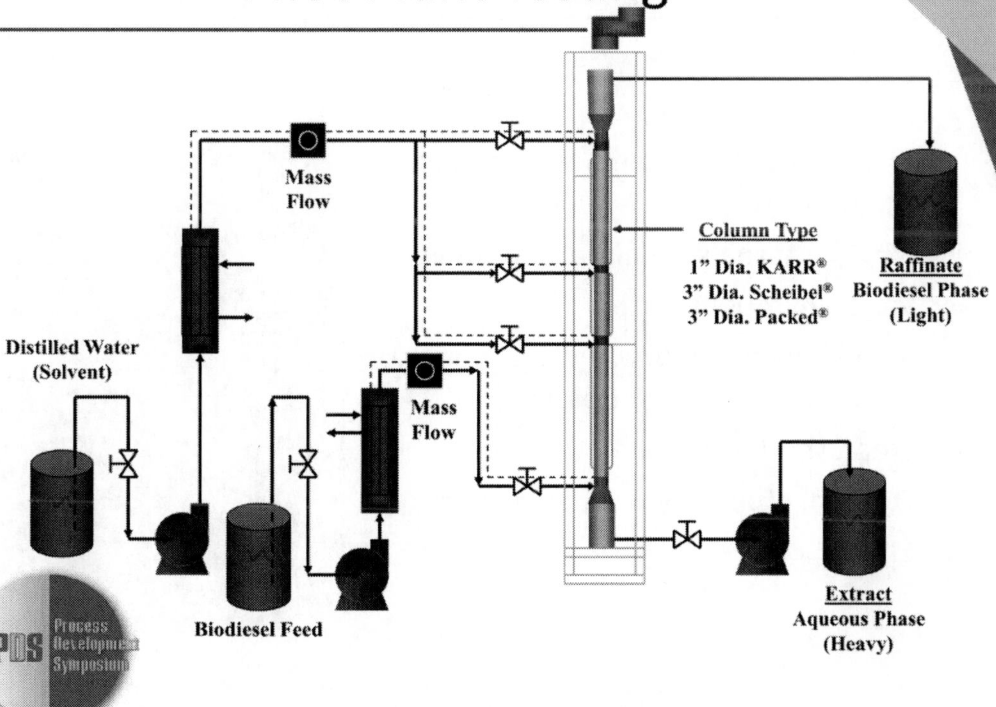

KARR® Column Testing

- Aqueous phase dispersed (interface at the bottom)
- PTFE plates and spacers- provided proper "wetting" characteristics
- Variable plate spacing
- Distinct disperse phase drops and clear, sharp interface
- Optimal efficiency= 1.2 stgs/meter

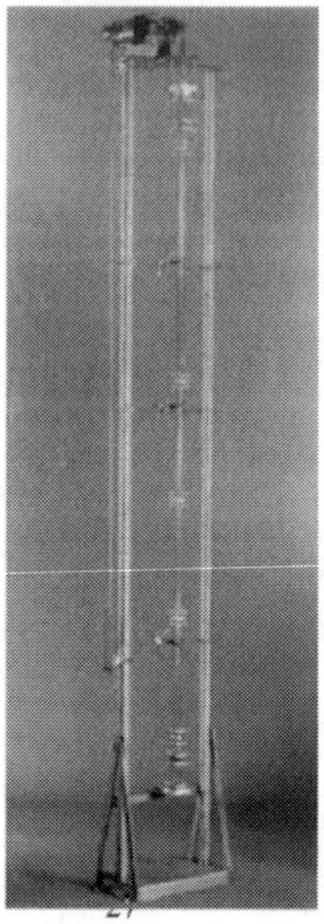

2016 Process Development Symposium | Providence, RI | June 7-9, 2016

SCHEIBEL® Column Testing

- Aqueous phase dispersed (interface at the bottom)
- 316SS internals- minimal coalescing
- Variable impeller sizes
- Smaller disperse phase drops and more hold-up than KARR® Column
- Clear, sharp interface
- Optimal efficiency = 5.1 stgs/meter

2016 Process Development Symposium | Providence, RI | June 7-9, 2016

Packed Column Testing

SMVP Packing- 316SS internals

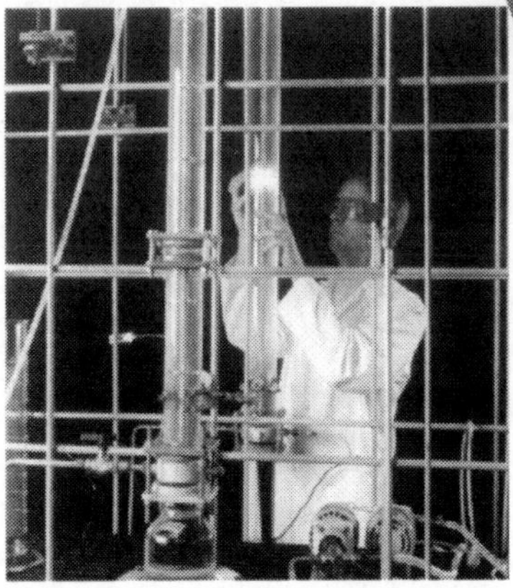

Organic Phase Dispersed

- Poor phase dispersion and hold-up. No clear interface
- Low capacity- sensitive to flooding

Aqueous Phase Dispersed

- Large dispersed phase drops, and low hold-up
- No clear interface- emulsion filling lower disengaging chamber

Efficiency Results from Pilot Testing

	Capacity (m^3/m^2hr)	Max Efficiency (Stages/Meter)
KARR®	20.3	1.2
	24.4	1.1
	28.5	1.0
SCHEIBEL®	16.3	5.1
PACKED	24.4	0.4
	28.5	0.3

Stichlmair Type Plot

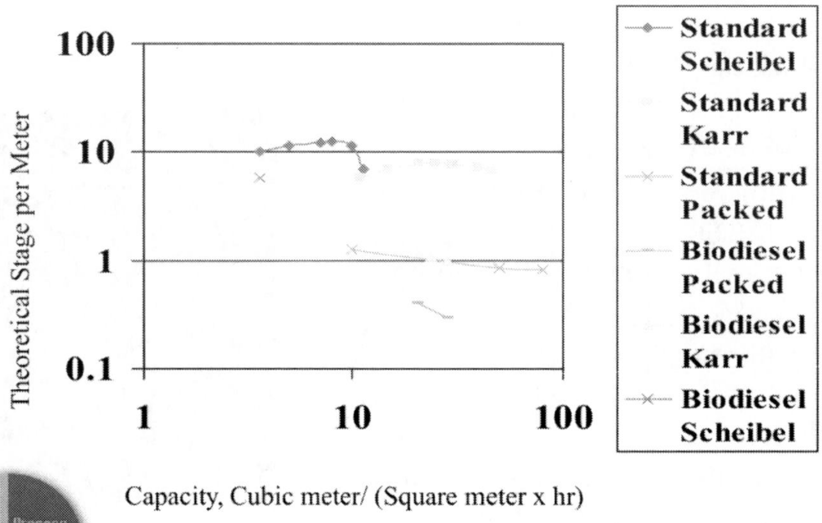

PDS Process Development Symposium

Extraction Columns in Series vs Distillation of Solvent Recovery

KMPS Koch Modular Process Systems, LLC.

MODULAR PROCESS SYSTEMS | DISTILLATION | LIQUID EXTRACTION | ENGINEERING | PILOT TESTING

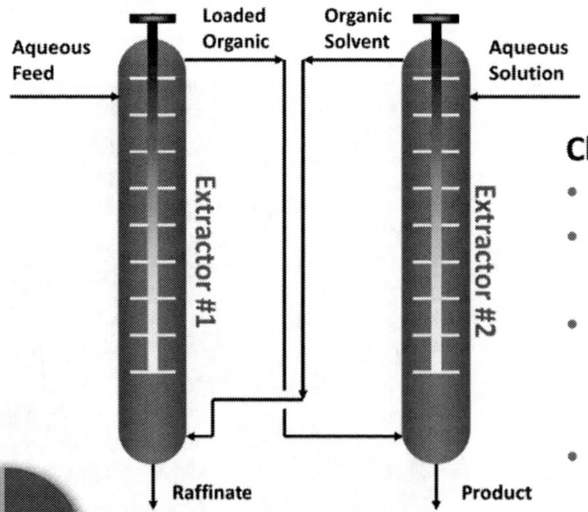

Characteristics:

- Solvent is recovered by
- stripping or back-extraction
- No distillation so water is conserved since no cooling water is required
- Low energy requirement

PDS Process Development Symposium

www.modularprocess.com
Telephone: (201) 267-8670

Nickel Recovery and Purification by Solvent Extraction

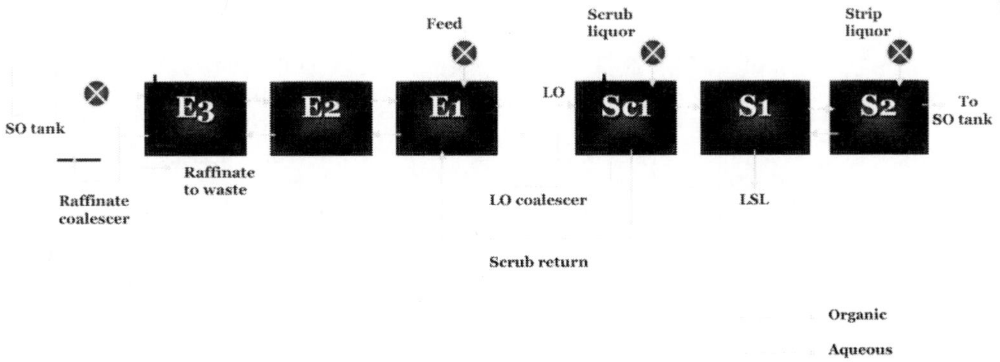

Cobalt and Nickel Extraction Mixer-Settlers in Series

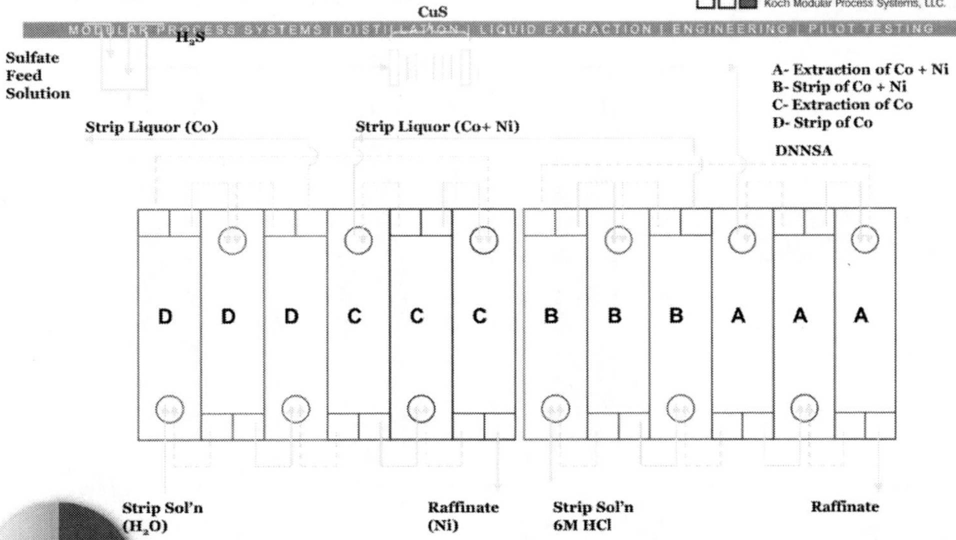

Mixer / Settlers for Metals

KMPS Koch Modular Process Systems, LLC.

MODULAR PROCESS SYSTEMS | DISTILLATION | LIQUID EXTRACTION | ENGINEERING | PILOT TESTING

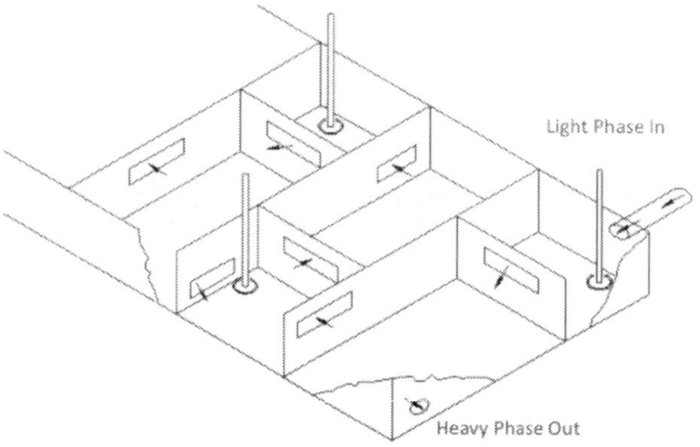

Characteristics

- Handle very high flowrates
- Good for processes with relatively slow reactions (residence time required)
- Provide intense mixing to promote mass transfer
- Require large amount of floor space
- Suitable when few theoretical stages required
- Large solvent inventory (and losses)

Phosphoric Acid Extraction

KMPS Koch Modular Process Systems, LLC.

MODULAR PROCESS SYSTEMS | DISTILLATION | LIQUID EXTRACTION | ENGINEERING | PILOT TESTING

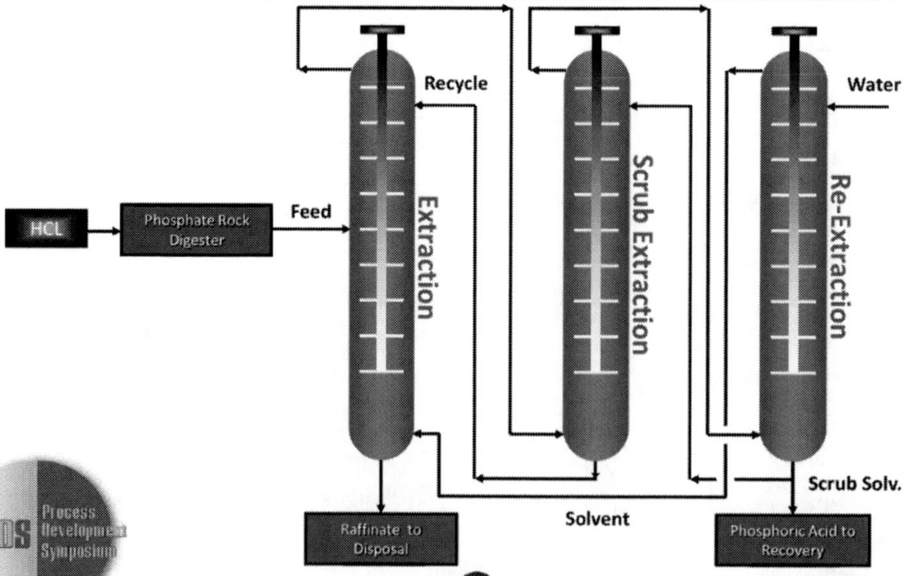

Recovery of Caprolactam

KMPS Koch Modular Process Systems, LLC.

MODULAR PROCESS SYSTEMS | DISTILLATION | LIQUID EXTRACTION | ENGINEERING | PILOT TESTING

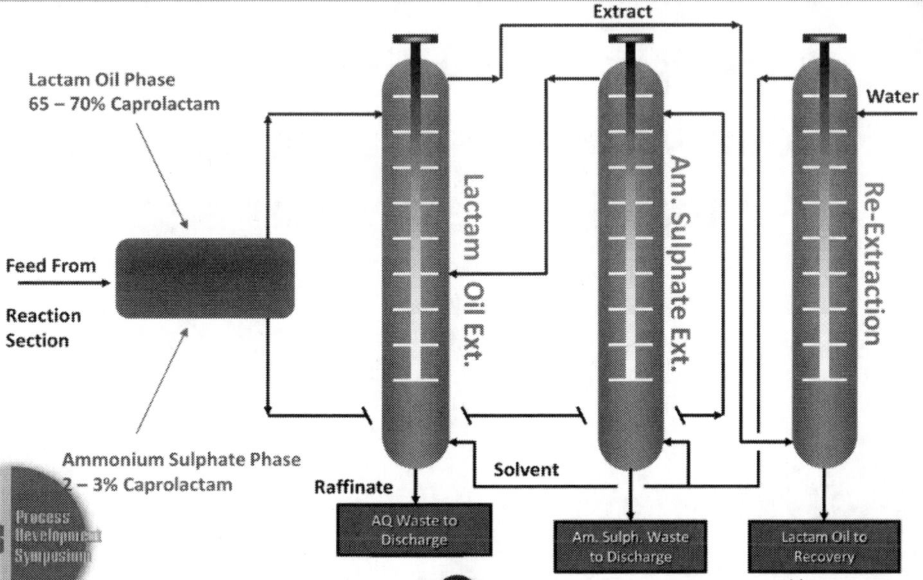

PDS Process Development Symposium

Thank You!

KMPS Koch Modular Process Systems, LLC.

MODULAR PROCESS SYSTEMS | DISTILLATION | LIQUID EXTRACTION | ENGINEERING | PILOT TESTING

Questions?

PDS Process Development Symposium

Design and Optimization of Recovery and Purification Processes for Biobased Materials

Erik Gommeren, PhD
DuPont Engineering Research and Technology

Process Development Symposium

2016 Process Development Symposium | Providence, RI| June 7-9, 2016

DUPONT®

DuPont Industrial BioSciences

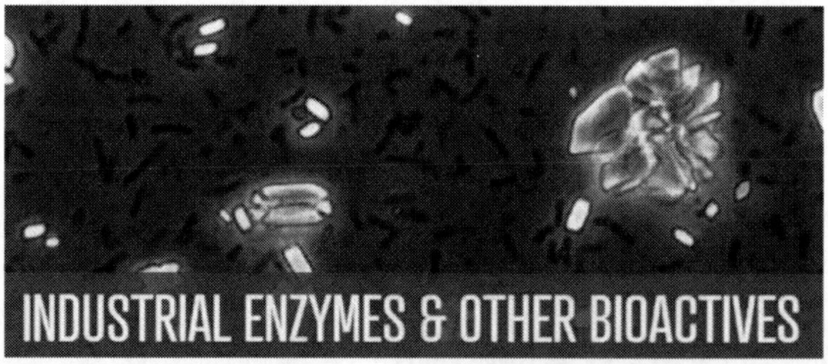

2016 Process Development Symposium | Providence, RI| June 7-9, 2016

DUPONT®

Biotech is Downstream Process Intensive

- Product recovery
- Product concentration
- Wastewater treatment (and water reuse)

} 50-70% of total costs*

Water Use : Product Produced =
40:1

Improved cost and productivity critical for competitiveness and sustainability

3

*The European Association for Bioindustries (2006)

Challenges in BioSeparation

Food and Biotech Products

Polysaccharides
$\eta = 10^{-3} - 10^{0}$ Pas
Microstructure

Solids

Crystals

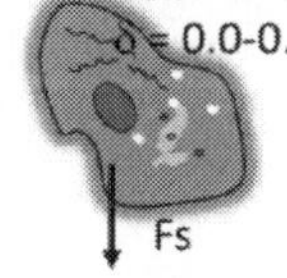

Cell = 1-10 μm
ρ = 0.0-0.1 kg/l
Fs

Proteins

Small molecules

Demands / Unmet Needs

- No residual solvents / Limited additives
- Functionality / Bio-availability / Activity
- Consumer acceptance (audits)
- Regulatory (FDA, EPA, DA)
- Sterility, cleanliness

Behavior / Properties

Denaturation and degradation

Compressible solids

Fouling

After K. Koller

4

Cost of BioSeparation

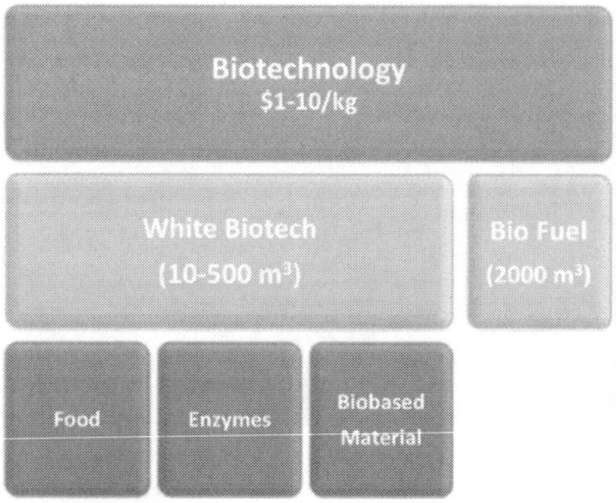

Product Type	Separation Cost (%)
Solvents – (bio fuels, bio polymers)	10-20
Live Cells & Crude cellular extracts	20-30
Organics acids	30-40
Vitamins, sweeteners, amino acids	30-40
Gums and polymers	40-50
Antibiotics	20-60
Industrial enzymes	25-65
Non-recombinant therapeutic proteins	50-70
r-DNA products	60-80
Monoclonal antibodies	50-70
Nucleic acid based products	60-80
Plasma proteins	70-80

Sources: Modified after R. Ghosh "Principles of Bioseparation Engineering". Additional inputs from E. Miller, B. Vrana & S. Kinsella

Demands for BioSeparation

- High throughput, High product yield
  - Large-scale
  - Continuously
- High purity
  - Selective
- Easy scale-up & no-scale limitations
  - Adjustable
- Low costs
  - Reduction in number of process steps
  - Pre-concentration with a simple separation device

Engineering Approach

Guidance and Methods

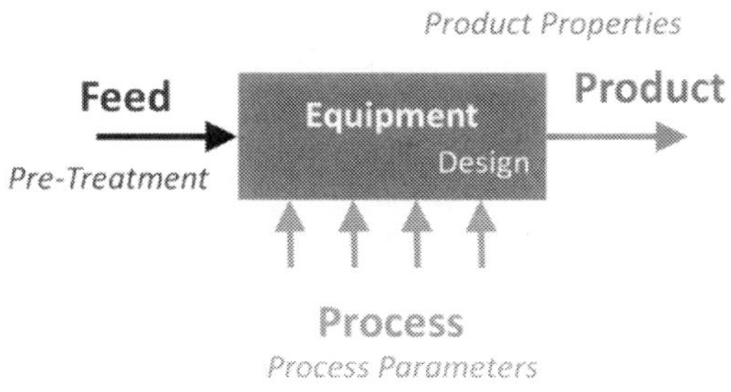

Separating the *Effects*

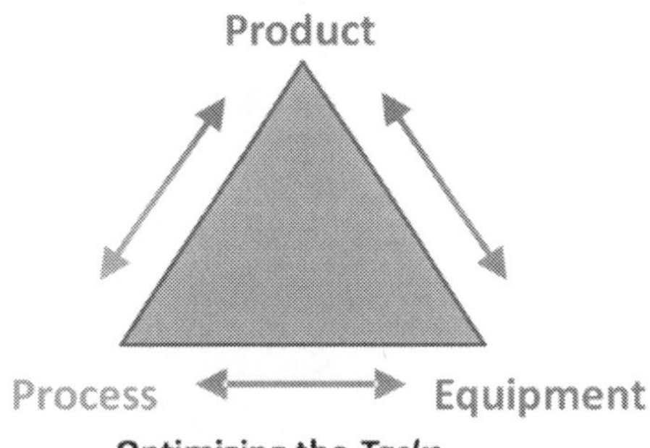

Optimizing the *Tasks*

Understanding the Product Properties
Optimizing the Process Parameters
Selecting the correct Equipment

After K. Keller

2016 Process Development Symposium | Providence, RI| June 7-9, 2016

7

Separation Equipment Options

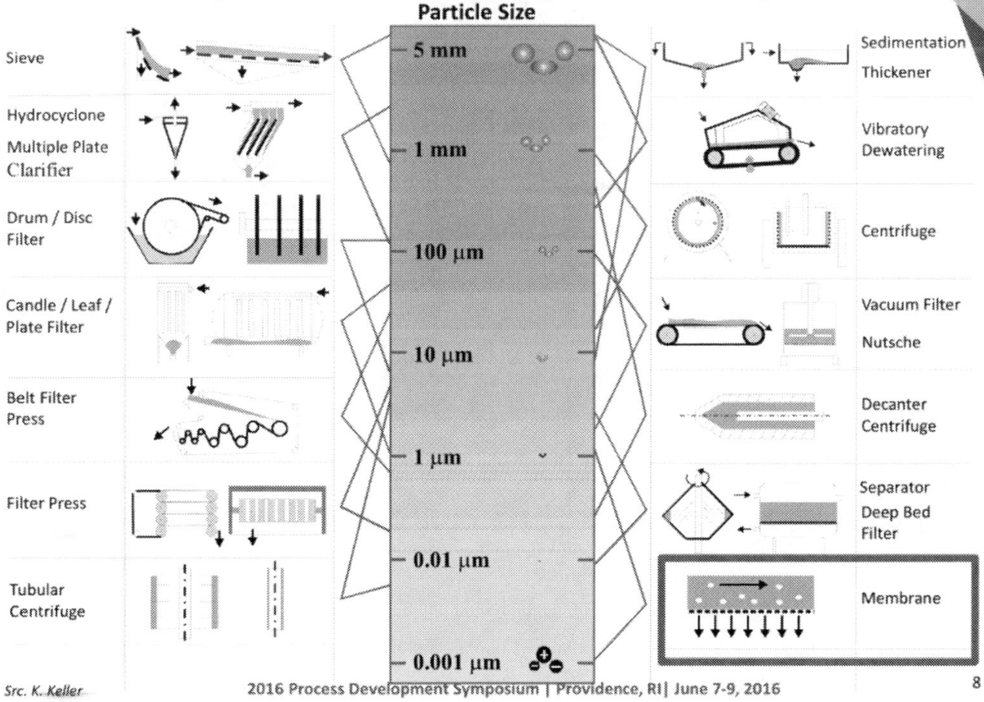

Src. K. Keller

2016 Process Development Symposium | Providence, RI| June 7-9, 2016

8

Membrane Applications in Food and BioTech

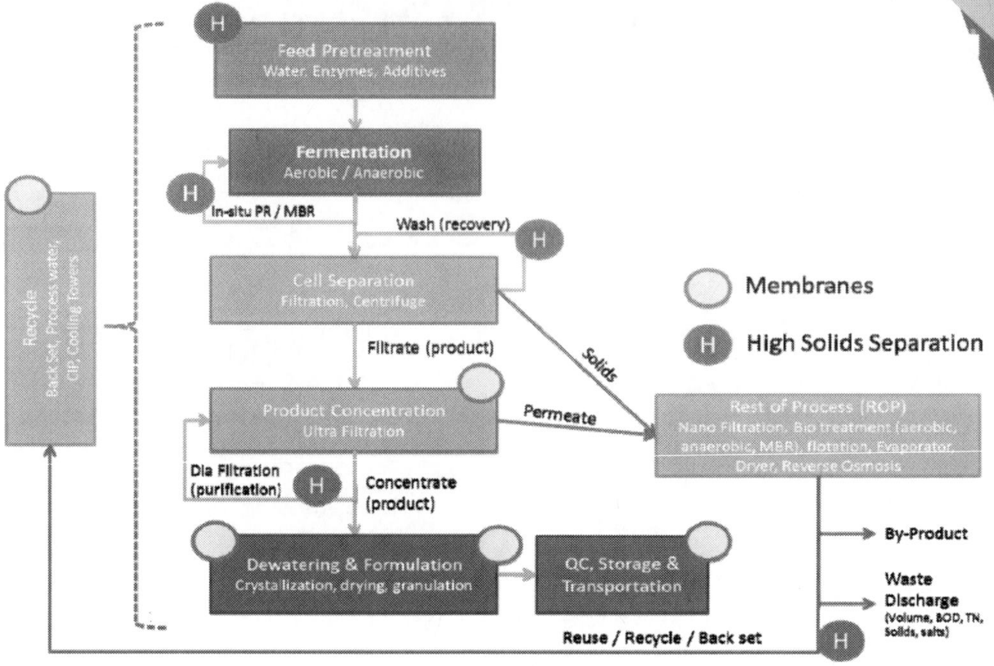

Filtration Modes

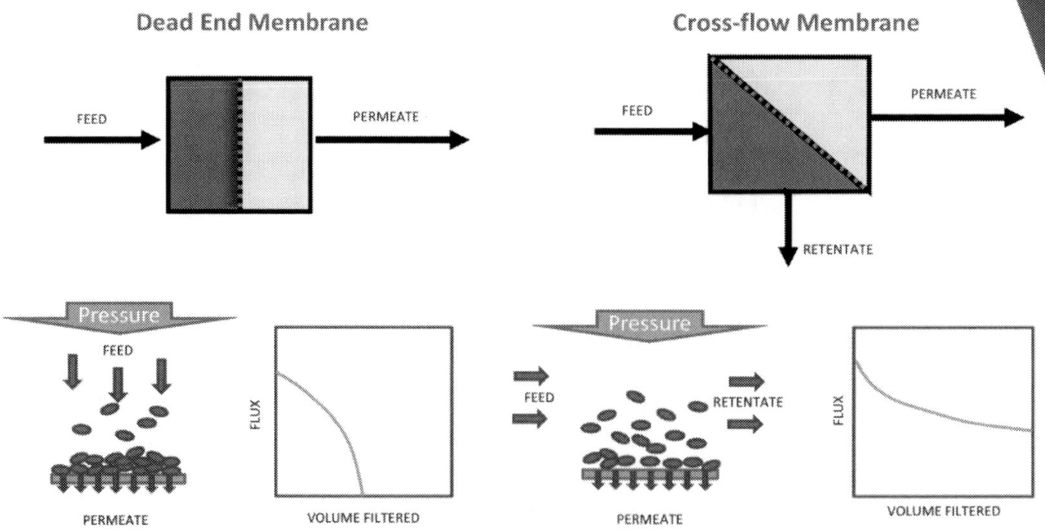

PERMEATE = FILTRATE RETENTATE = CONCENTRATE

Membrane Retention

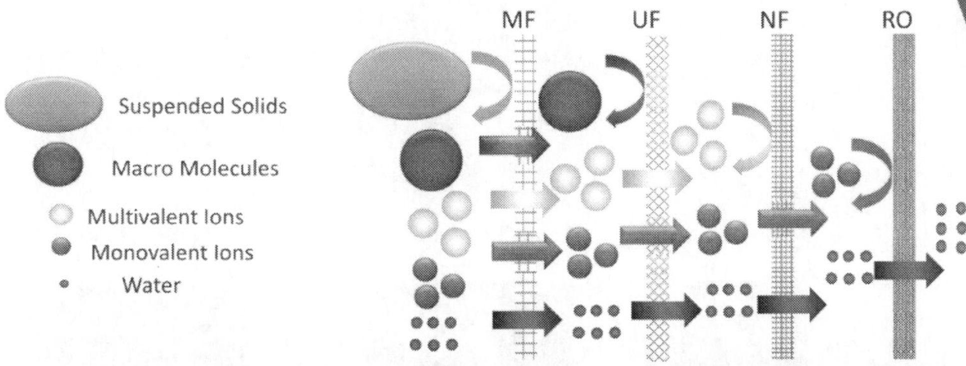

Best Practices:

- Use sufficiently high retention to meet yield goal.
- To retain: Select pore size (micron, MWCO) 1/3rd to 1/5th of the size of product.
- For pass through: Select pores size 3 to 5 times larger than product.
- To Separate: > 5-fold size difference between components.

11

Membrane Technologies

Advantages

- No phase change
- Energy efficient
- Higher quality product
- Low temperature (no heat for sterilization or evaporation). No product degradation.
- Environmentally friendly
- Easy to scale-up

Limitations

- Long-term reliability not proven
- Excessive pretreatment
- Concentration polarization and membrane fouling
- Replacement of membranes

Membrane Formats

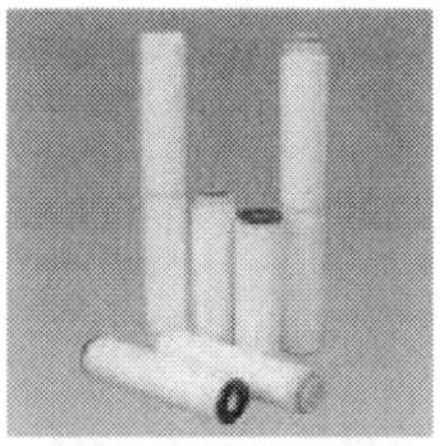

Pleated Cartridge

Hollow Fiber

Plate & Frame

Monolithic

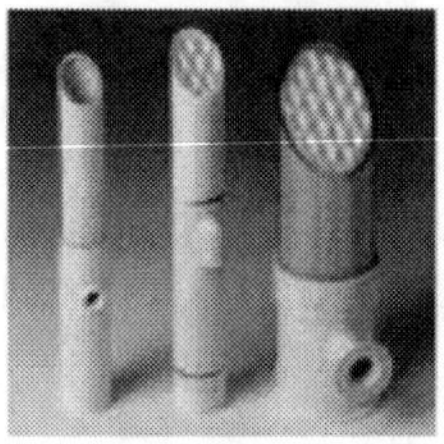

Tubular

Spiral Wound

Membrane Material Compatibility

CHEMICAL NAME	GROUP	Nylon	Acrylic	PVDF	PES	PP	PVC	Glass, Silica	MCE	ABC	PET	PTFE	UHMWPE
Density [g/cm3]		1.14		1.78		0.91	1.38					2.18	
T_Melt [C]				165		160	80					342	130
Acetic acid, 10%	*ACIDS*	N	-	R	R	R	R	R	-	N	R	R	R
Acetic acid, 30%	*ACIDS*	N	-	R	R	R	R	R	-	N	L	R	R
Acetic acid, 90%	*ACIDS*	N	-	R	R	R	R	R	-	N	N	R	R
Acetic acid, glacial	*ACIDS*	N	N	R	R	R	R	R	N	N	N	R	R
Hydrochloric acid, 1N (3.3%)	*ACIDS*	N	-	R	R	R	R	R	-	N	R	R	R
Acetic anhydride	*ACIDS*												R
Hydrochloric acid, 6N (20%)	*ACIDS*	N	-	R	R	R	R	R	-	N	R	R	R
Hydrochloric acid, conc. (35%)	*ACIDS*	N	N	R	R	R	R	R	N	N	R	R	R
Nitric acid, 6N (27%)	*ACIDS*	N	-	R	-	R	R	-	-	N	L	R	R
Nitric acid, conc. (67%)	*ACIDS*	N	N	R	N	R	R	L	N	N	N	R	R
Sulfuric acid, 6N (16%)	*ACIDS*	N	-	-	-	R	R	R	-	N	L	R	R
Sulfuric acid, conc. (96%)	*ACIDS*	N	N	N	N	R	R	R	N	N	N	R	R
Amyl alcohol	*ALCOHOLS*	R	R	R	N	R	R	R	R	R	L	R	R
Benzyl alcohol	*ALCOHOLS*	L	R	R	N	R	R	N	N	R	N	R	R
Allyl alcohol	*ALCOHOLS*												R
Butanol	*ALCOHOLS*	R	R	R	R	R	R	R	L	R	L	R	R
Ethanol	*ALCOHOLS*	R	R	R	R	R	R	R	N	R	R	R	R
Isopropanol	*ALCOHOLS*	L	R	R	R	R	R	R	R	R	L	R	R
Methanol	*ALCOHOLS*	R	R	R	R	R	R	R	N	-	L	R	R
Benzene	***AROMATICS***	**L**	**R**	**R**	**R**	**R**	**N**	**R**	**R**	**R**	**R**	**R**	**L**
Toluene	***AROMATICS***	**L**	**R**	**R**	**R**	**R**	**N**	**R**	**R**	**R**	**L**	**R**	**L**
Xylene	***AROMATICS***	**L**	**R**	**R**	**R**	**R**	**N**	**R**	**R**	**R**	**L**	**R**	**L**
Ammonium hydroxide, 3N (5.7%)	*BASES*	L	R	N	R	R	R	R	-	N	R	R	R
Ammonium hydroxide, 6N (11.4%)	*BASES*	L	R	N	R	R	R	R	-	N	L	R	R
Potassium hydroxide, 3N (15%)	*BASES*	N	R	N	R	R	R	N	N	N	R	R	R
Sodium hydroxide, 3N (11%)	*BASES*	L	R	N	R	R	R	-	-	N	R	R	R
Sodium hydroxide, 6N (22%)	*BASES*	N	R	N	R	R	R	-	-	N	L	R	R
Amyl acetate	*ESTERS & GLYCOLS*	R	R	R	R	R	N	R	N	R	L	R	
Butyl acetate	***ESTERS & GLYCOLS***	**R**	**L**	**R**	**R**	**R**	**N**	**R**	**N**	**R**	**R**	**R**	
Cellosolve acetate	*ESTERS & GLYCOLS*	R	R	L	R	R	N	R	N	R	-	R	
Ethyl acetate	*ESTERS & GLYCOLS*	R	R	R	N	R	N	R	N	-	R	R	
Ethyl ether	*ESTERS & GLYCOLS*	R	R	R	R	R	N	R	N	N	N	R	

Cost Benefit Analyses – Alternative Membrane Technologies

Charters	Description	Hi Solids / viscosity	Low Fouling & Cleanability	Cost (Capex / Opex)	Enzyme Yield Recovery	Low Technology Risk
	Spiral Wound Cross Flow	*	*	***	***	***
	Plate & Frame Cross Flow	***	***	**	**	**
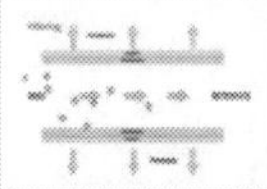	Tubular Cross Flow	***	***	**	**	**
	Dynamic filtration	****	***	*	*	*
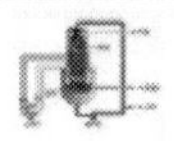	Evaporator	*	*	*	*	*

Spiral Wound Elements

- Benefits: compactness, ease and economy of manufacture, and ease of replacement.
- Industry standard. Many manufacturers.

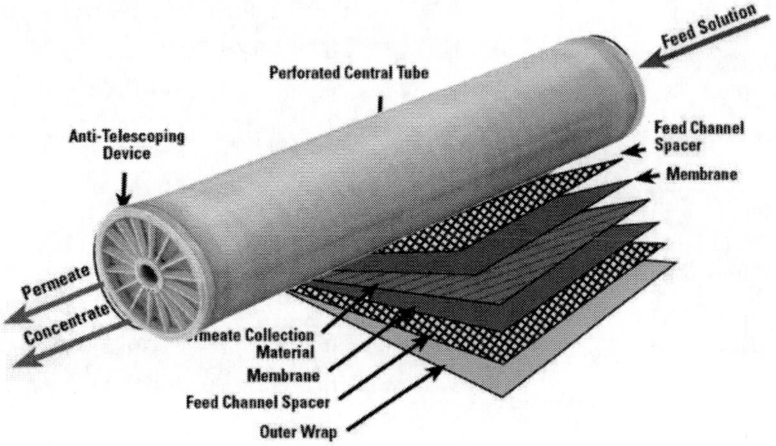

Elements sizes

Code	
2517	= 2.5” diameter x 17” length
4040	= 4” x 40” length
8040	= 8” diameter x 40” length
12040	= 12” diameter x 40” length

Note: Elements from different vendors are (often) interchangeable

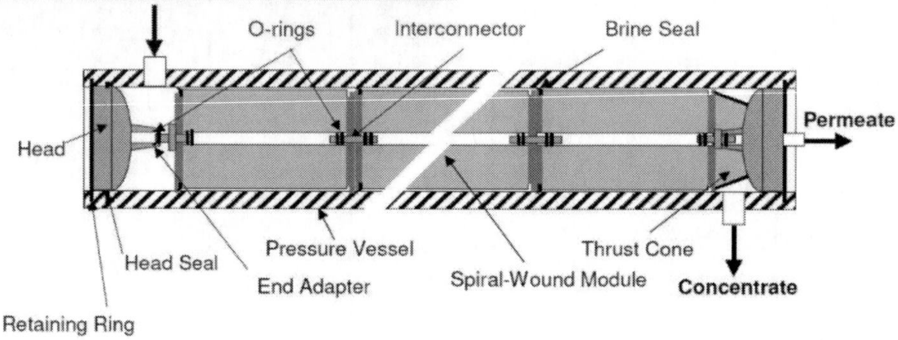

17

BioSeparation – Technology Selection and Design

1. Assessment of Needs
2. Initial Evaluation
3. Piloting and Modeling
4. Scale-up & Final Design
5. System Fabrication

- Full Scale operation: Commissioning and Start-up
- Holistic and Integrated Approach

2016 Process Development Symposium | Providence, RI| June 7-9, 2016

18

Product Properties

- State of Matter: Form in which bio-product is present in fermentation broth
  - Dissolved
  - Solid (eg. crystal)
  - Separate liquid phase
  - As a gas
- Determines opportunities for recovery and purification

Characterization

Concentration

Densities, Viscosities

Particle size distribution

General feasibility of Filtration

Influenced by customer demands or particle properties

Empirical/Experience

Benchtop Membrane Systems – Multi Format

Bench top membrane unit

1. Nano Filtration and Reverse Osmosis

- Separation of salts and sugars

2. Ultra Filtration (5-100kD)
- For separation of larger molecules (proteins, poly sacch.)

3. Micro Filtration (0.1 micron – 5 micron)
- Micro organism removal / cell harvesting / clarification
  - Particulate removal. Feed pretreatment.

Membrane formats

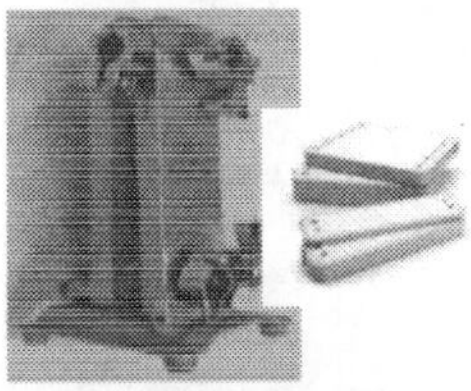

Cassettes

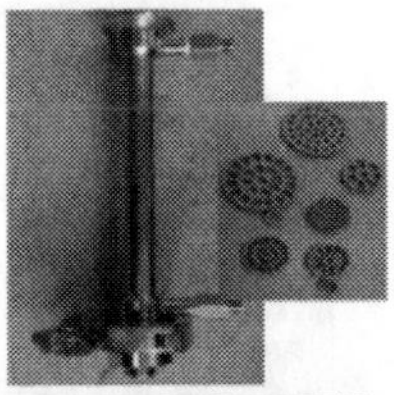

Tubular

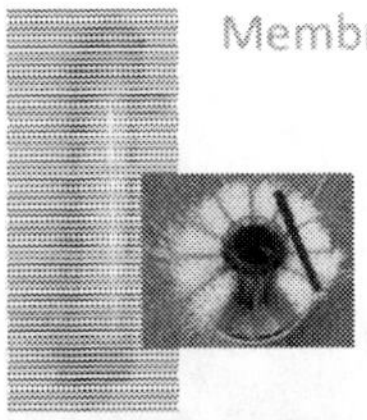

Hollow Fiber

Flat Sheet

Trans-Membrane Pressure (TMP)

Trans-membrane Pressure

$$TMP = \left(\frac{P_F + P_R}{2}\right) - P_P$$

Differential Pressure

$$\Delta P = P_F - P_R$$

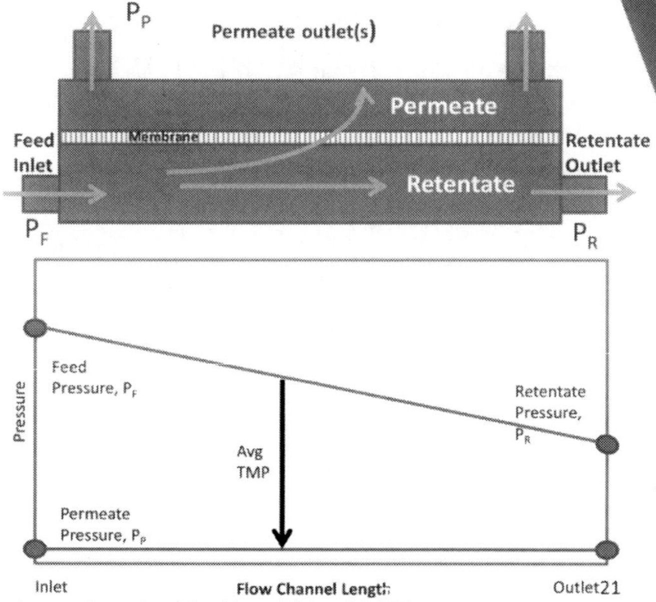

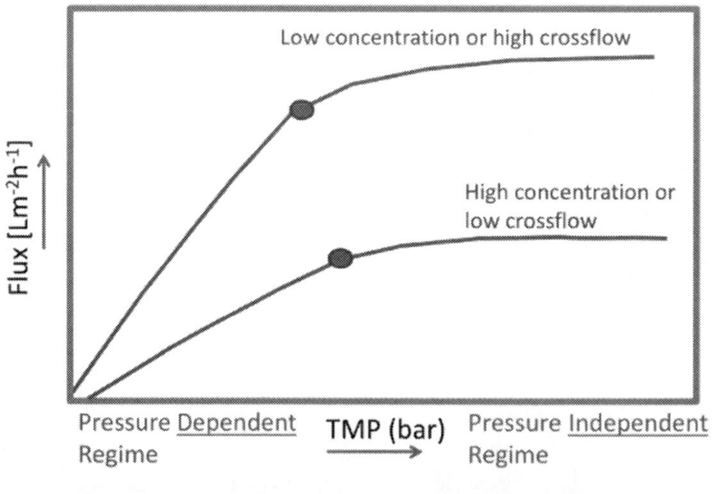

Optimal TMP at knee of curve. (Nearly highest flux)
Avoids excessive pressure and reaching high wall concentrations (polarization).

Plant Issues: Membrane Fouling

Bio growth in tanks and pipes hits membranes

- Despite Ultra filtration, Nano membranes can foul quickly
  - Due to growth between systems (in tanks and pipes)
  - Significantly reduces membrane life
  - Difficult to manage (limitations: membrane chemistry, food grade, ect.)

warping

Origins of Fouling	Fouling Control
Scaling: substances exceeding their solubility product	Operate below solubility limit, pretreatment: reduce pH to 4-6, low recovery, and anti-scalants. Pre-oxidation of metals.
Deposition of colloidal matter or dispersed fines	Pretreatment using filtration, microfiltration (MF) or Ultrafiltration (UF)
Organic fouling	Pretreatment using filtration, MF, UF, ion exchange, ozone, enhanced coagulation or carbon adsorption
Biofouling: colonization by bacteria	Hydrodynamics, operation below critical flux, chemical cleaning, pretreatment: disinfection or UF, MF, Hydrodynamics, operation below critical flux, chemical cleaning

Sources: Keith Riegel, A. Schuler

Membrane Fouling

- Definition
  - **Reversible** or **irreversible** deposition of material on membrane surface.
  - Decreases permeate flux and alters retention characteristics
- Consequences
  - ⇓ Flux (⇑ membrane area/ ⇑ processing time)
  - ⇑ Cleaning
  - ⇑ Membrane damage
  - Altered retention/permeability
  - ⇓ Separating ability
- Operating parameters affecting fouling rate
  1. Feed stream velocity
  2. Process Temperature
  3. Pressure

Multiple mechanisms

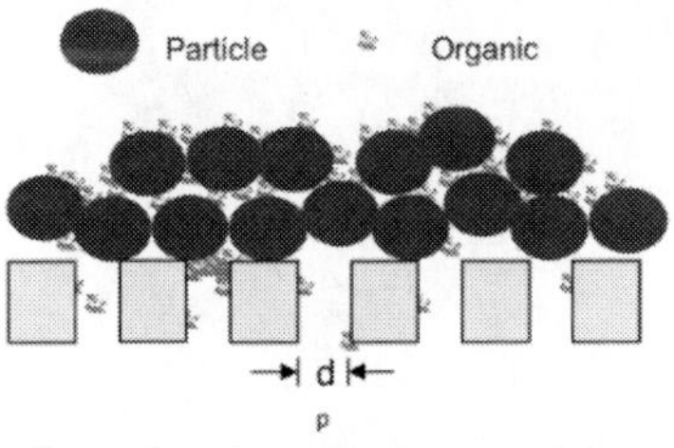

Organic adsorption on particle surface and membrane surface and/or pores.

3 D images comparison

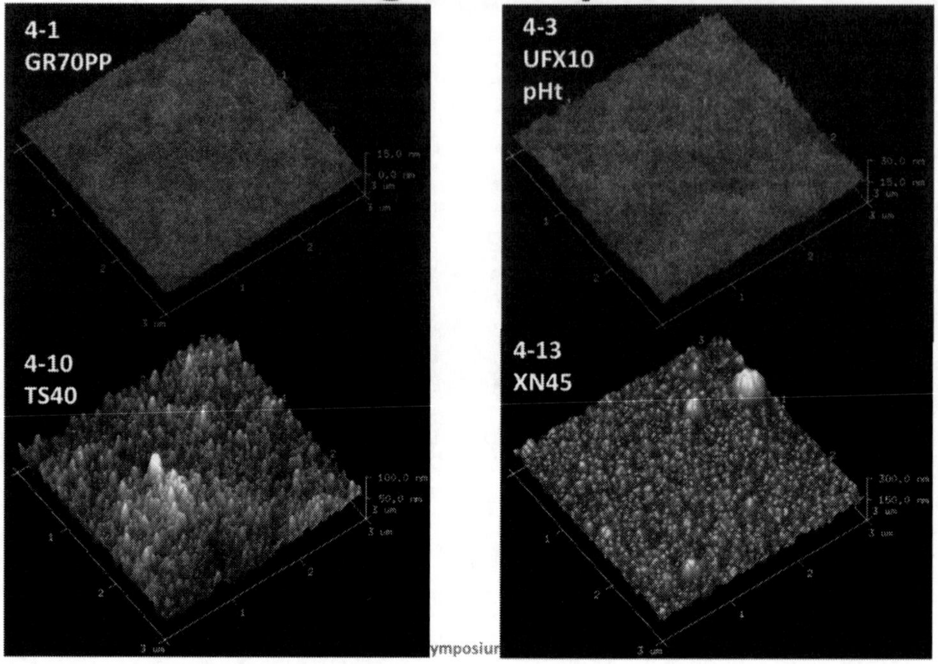

ymposiu

25

Membrane surface functionality

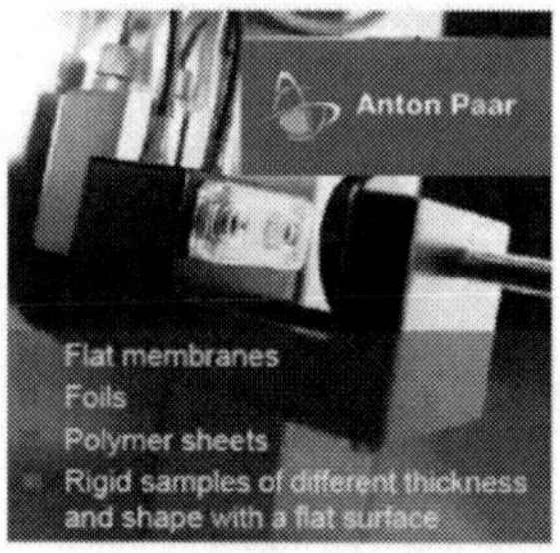

Membrane Property Characterization

- Develop method for surface electrokinetic analyses (**Anton Paar SurPASS**) to better understand effects of pH and ionic strength, and interaction of charged molecules/particles with membrane.
- Characterize **membrane fouling** and CIP chemical effectiveness .

Measures **streaming current** or **potential** to determine the **zeta potential** of solid **surfaces** using Helmholtz-Smoluchowski equation

$$\zeta = \frac{dI}{dp} \times \frac{\eta}{\varepsilon \times \varepsilon_0} \times \frac{L}{A}$$

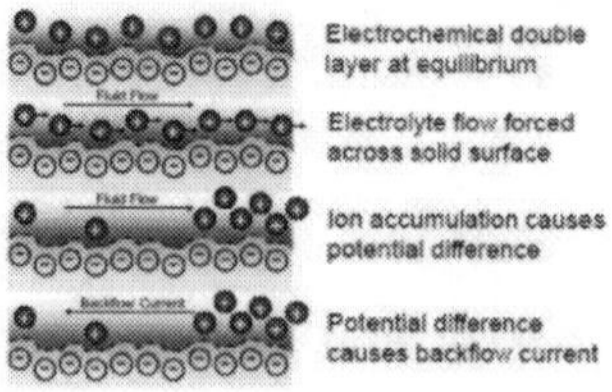

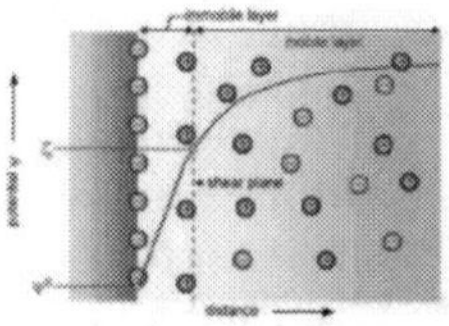

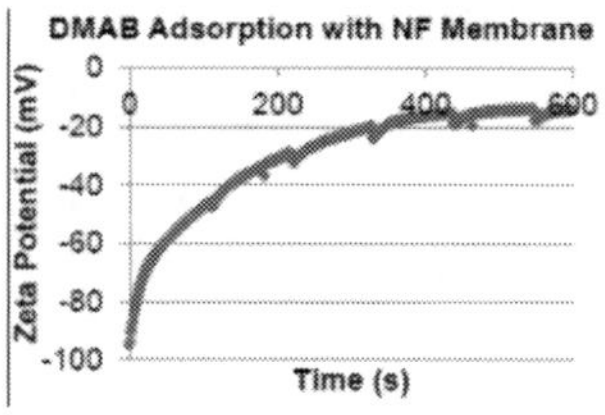

- Evaluate the kinetics and equilibrium of adsorption and desorption
- Can be applied to understand fouling processes and CIP cycles

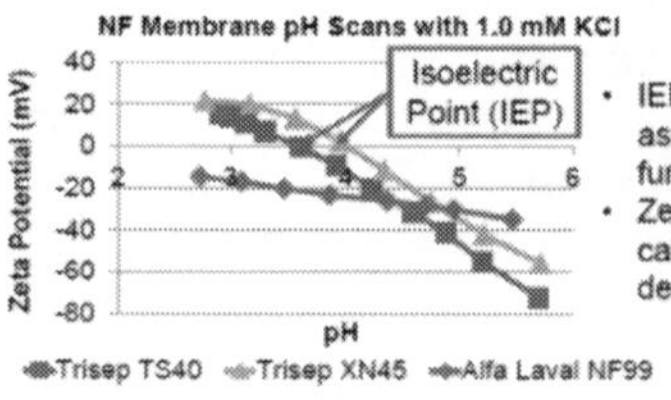

- IEP comparison allows assessment of potential functional surface groups
- Zeta potential used to calculate surface energy density

ium

D. Eicholtz

Membrane System Fabrication

Engineering Assistance:
Functional design (elements, staging)
P&ID (sensors, valves, pumps)
Control strategy / automation (DCS)
Safety (PRV, Alarms)
Factory Testing (FAT plan)

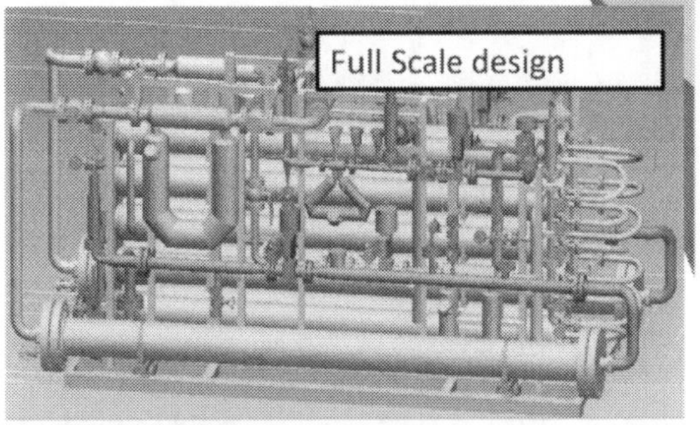

Membrane Cost Estimator

- Based on pilot scale testing.
  - Flux and selectivity
  - Fouling potential / cleaning
  - Optimal cross flow and membrane pressures
- CAPEX and OPEX

Pilot scale membrane skid

Application	Nano filtration system - HI SPOT		
Business			
Location	TBD		
Incoming Stream	Reactor filtrate		
Effluent Quality	NF		
Membrane Specifications			
Membrane Type			*NF-SW*
Incoming flow	Kg/hr		*111,000*
Retentate Conc. Factor	x		*5.0*
Tracking Solute	g/Kg		*77*
Membrane Flux	LM2H		*15*
Design Factor			*2*
Density	Kg/L		*1.05*
Labor	min/d		*360*
Membrane replacement	yrs		*5*
Membrane model	[select]		
Cost Estimates			
Capital Cost	$/m2	$	*300*
Membrane costs	$/m2/yr	$	*25*
Electricity usage	KWh/y/m2	$	*29*
Cleaning Chemicals	$/y/m2	$	*27*
Labor cost	$/hr	$	*50*
System Design			
Permeate Flow	Kg/hr		*88,800*
Retentate Flow	Kg/hr		*22,200*
Solute Concentration	g/Kg		*385*
Membrane Area needed	m2		*11,276*
Area per casette	m2/element		*30*
# of elements	#		*376*
Cost of Operations (Total)			
Total Capital (4 x bare eq)	$	$	*3,382,857*
Capital Depreciation	$/yr	$	*225,524*
Membrane costs	$/yr	$	*281,905*
Electricity usage	$/yr	$	*328,363*
Cleaning Chemicals	$/yr	$	*304,457*
Annaul labor cost	$/yr	$	*75,000*
Yearly Operational Cost	**$/yr**	**$**	***1,215,248***

BioSeparations – Opportunities

- **Improved Product Yield**
  - More efficient /fewer steps → sell more product with same resources & Reduced waste volumes
  - Product Stability: Process conditions.: temperature, shear, residence time, oxidation, O2 starvation
- **Holistic Process Design**
  - Understand implications of upstream / downstream.actions
  - Scientists / Engineers talk in early stages)
- **Understand Thermodynamics**
  - Compare against energy value of fuels, and value of final product.
  - Mechanical versus Thermal Water removal
- **Cleanability and Cleanliness**
  - Equipment design, surfaces, CIP systems, QC. Training
- **Robustness of Process**
  - "natural" variation in bio reaction. Flexibility (multiple products)
- **Continuous Plant Improvements**
  - Heat & Energy integration + Water usage integration
  - Process Integration: In-situ product / inhibitor removal. Fewer process steps (footprint, cost)
- **Include "Rest of Process"**
  - Waste water treatment, reuse, and by products. Consider during feasibility.
  - Leverage between multiple plants
- **Value from by-products**
  - Key to future competitiveness and sustainability

Fundamentals of 'Gas-Assisted' Separations for Dilute Aqueous Systems

Scurto, A. M., University of Kansas Chemical and Petroleum Engineering & Center for Environmentally Beneficial Catalysis

Originally presented on: 6/8/2016 11:45:00 - 12:15:00

The separation of dilute aqueous systems is a reoccurring challenge within industry, especially for processes with biomass or bio-renewable based feedstocks. Many processes involve energy-intensive separations such as distillation, liquid-liquid extraction followed by distillation, etc. which significantly affects costs and sustainability. As a potential alternative to thermal-based approaches, we introduce a method called "Gas-Assisted" Separations. Here, a gas is dissolved into the aqueous solution to induce the target solute from the aqueous phase into a new phase (liquid or solid) rich in that target solute. The systems can be decanted/filtered to recover the solute and the gas can be recovered with depressurization. These processes differ from supercritical fluid technology where the compressed gas/fluid is the solvent itself and require much higher pressures. This "salting-out" with an easily recoverable dissolved gas "antisolvent" (versus a solid salt or liquid, etc.), may have very different molecular mechanisms for each gas and solute combination. For instance, the use of gaseous CO2 would lead to both a decrease in pH and increase in the ionic strength which may have specific interactions with a given solute. In this presentation, the fundamental phase equilibrium will be overviewed for this vapor-liquid-liquid equilibrium (VLLE) behavior with temperature, pressure, and initial loading. Examples from literature will be used to illustrate the process.

sabic

PROCESS REVITALIZATION

2016 Process Development Symposium | Providence, RI | June 8, 2016

Greg Hemmer

CHEMISTRY THAT MATTERS™

sabic

Greg Hemmer, Chief Scientist

SABIC's Innovative Plastics Business – Technology & Innovation, High Performance Polymer, Global Process Technology

Josh Croll, Senior Manager

SABIC's Innovative Plastics Business – Technology & Innovation, High Performance Polymer, Global Process Technology

Evelyn Pearson, Senior Manager

SABIC's Innovative Plastics Business – Technology & Innovation, Core Polymer, Global Process Technology

PDS Process Development Symposium

No. 1

ASSET DESIGN

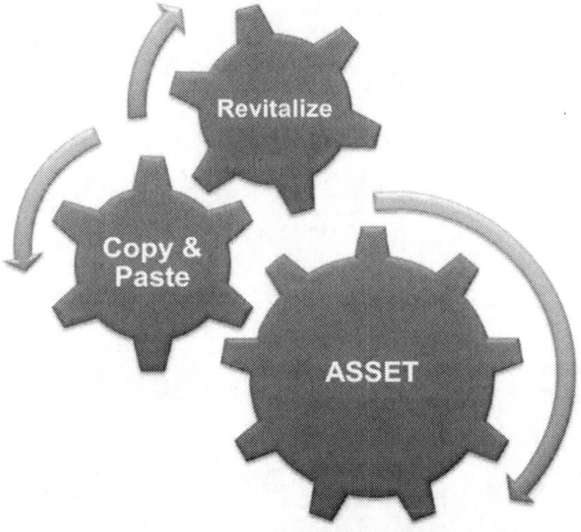

COPY AND PASTE APPROACH

+ SPEED
+ CERTAINTY OF RESULT
- HIGHER CAPEX & OPEX
- COMPETITIVE ADVANTAGE

REVITALIZATION

- SPEED
- CERTAINTY OF RESULT
+ HIGHER CAPEX & OPEX
+ COMPETITIVE ADVANTAGE

PROCESS REVITALIZATION DEPENDS ON TIME REQUIRED TO DEVELOP AND ASSESS CHANGE AND THE ACHIEVABLE / TANGIBLE BENEFITS

No. 2

PROCESS IMPROVEMENT ROADMAP

TIME

LEVEL 1
Optimize

Minimal changes in existing chemistries

Simplify and right sizing / scaling of existing unit operations

External manufacturing as a source of supply vs in-house synthesis

LEVEL 2
Modernize

Modest changes in existing chemistries

Scaling and intensification of existing unit operations

Batch to continuous transformations

External manufacturing as a development partner

LEVEL 3
Next Gen

Alternative reaction pathways / chemistries and solvents

Alternative unit operations and advanced intensifications

Extensive pilot and semiworks required

External manufacturing as semiworks

IDEATION PROCESSES

FOUNDATIONAL INFORMATION

No. 3

IDEATION - GENERATION

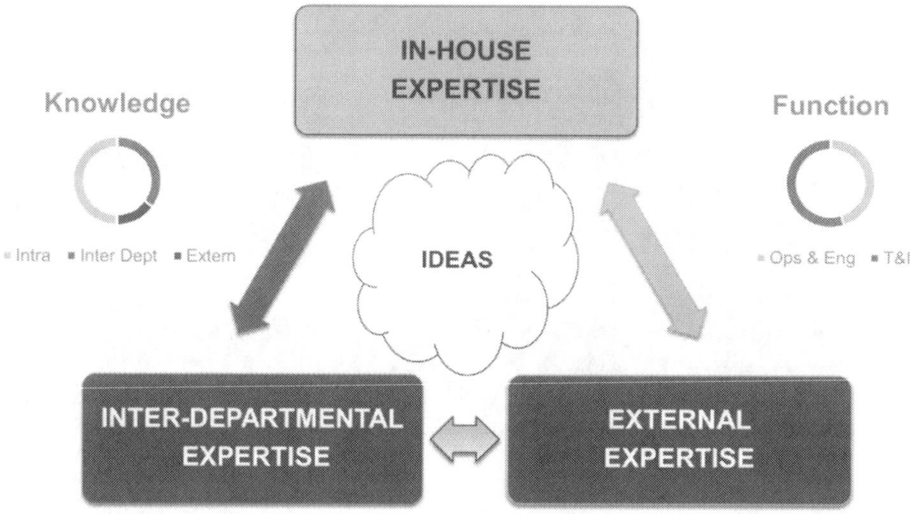

Actively seek knowledge and expertise to generate new ideas

No. 4

IDEATION - SELECTION

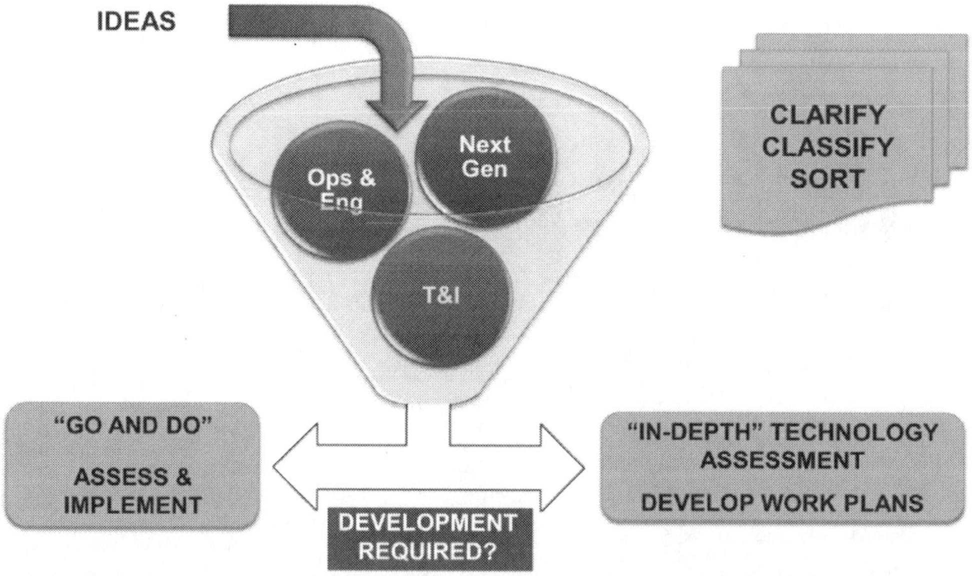

No. 5

FOUNDATIONAL INFORMATION

Documentation

- Process narrative
- Equipment & process details
- Operating strategies
- Critical to design parameters
- Material of construction

Properties & Kinetics

- Time vs temperature sensitivity data
- Liquid-liquid partition coefficients
- Crystallization kinetics
- Chemical reaction kinetics

Modeling

- Deeper insight regarding ex. unit ops
- Model-centric process development allows virtual experimentation

Benefits

Questioning the Status Quo
Uncovering inconsistencies
Identifying
- Competency needs
- Equipment design
- Operational protocols
- Control logic

Discovery of hidden factories

Build firm foundation for future change efforts

Identify no-cost opportunities to increase capacity in the existing process

No. 6

LEVEL 1 (OPTIMIZATION) EXAMPLE #1

Two-step to One-Step Acid Recovery

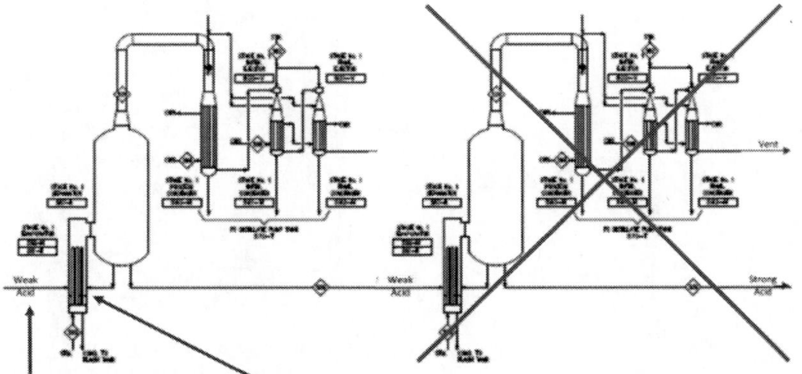

<u>Process Change</u> - concentration can be increased by various upstream modifications

<u>Equipment Change</u> – exchanger tubing requires high alloy material of construction – advances in fabrication allow scaling versus original installation

No. 7

LEVEL 1 (OPTIMIZATION) EXAMPLE #2

Two-step to One-Step Monomer Dissolution Processes

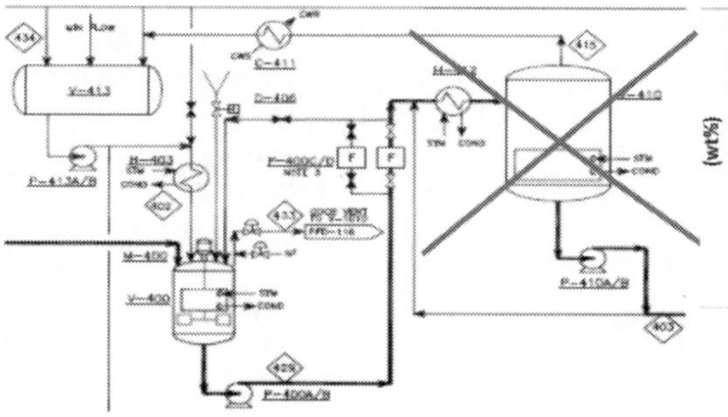

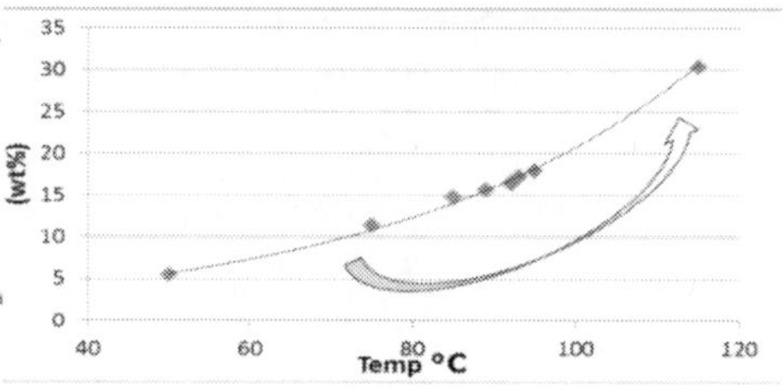

Improving foundational information and questioning status quo drives intensification

No. 8

LEVEL 2 (MODERNIZATION) EXAMPLE #1

Acid Concentration Column Intensification

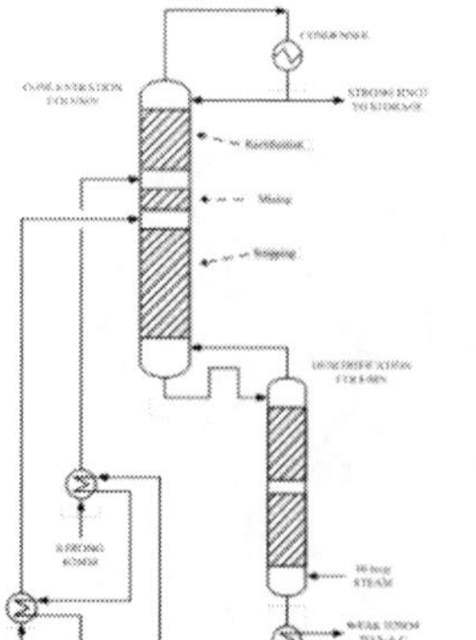

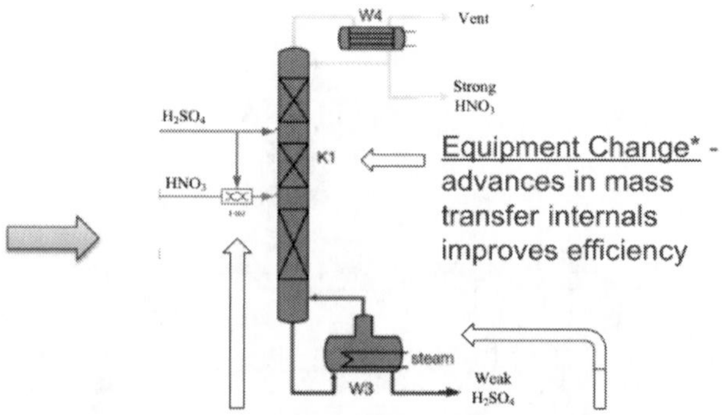

Leverage external expertise with proven commercialization experience to obtain readily available off the shelf improvement

* Schematic courtesy of De Dietrich Process Systems, Inc.

No. 9

LEVEL 2 (MODERNIZATION) EXAMPLE #2

Solvent Distillation Intensification

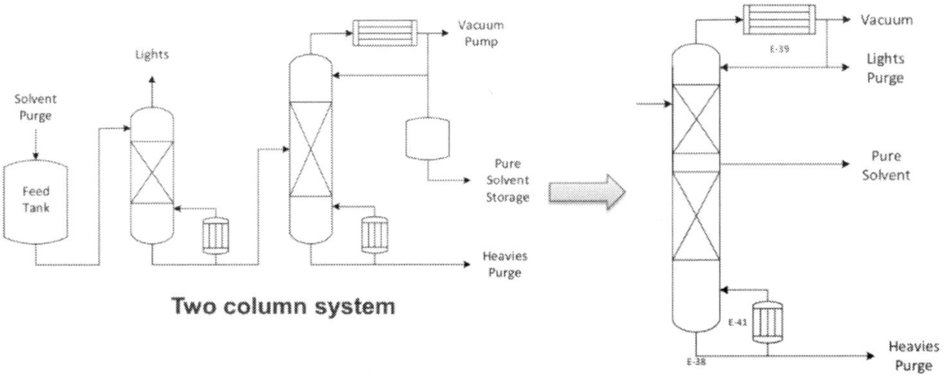

Two column system

Single column with side draw

➤ Solid Foundation + Expertise = Low Risk Intensification

LEVEL 3 (NEXT GEN) EXAMPLE #1

NOx Capture

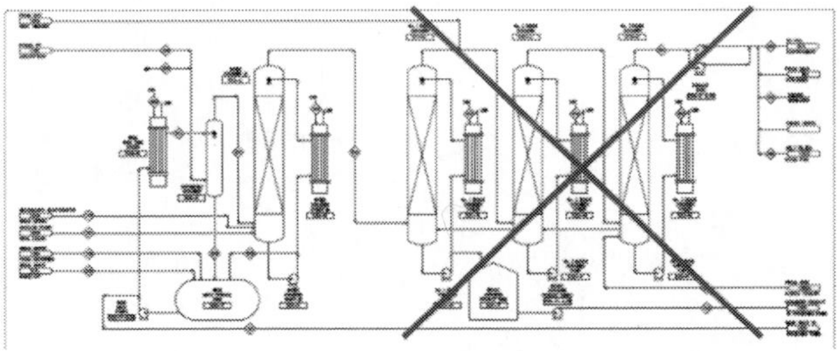

Gas absorption via multiple towers using two absorbents - early 80's BACT

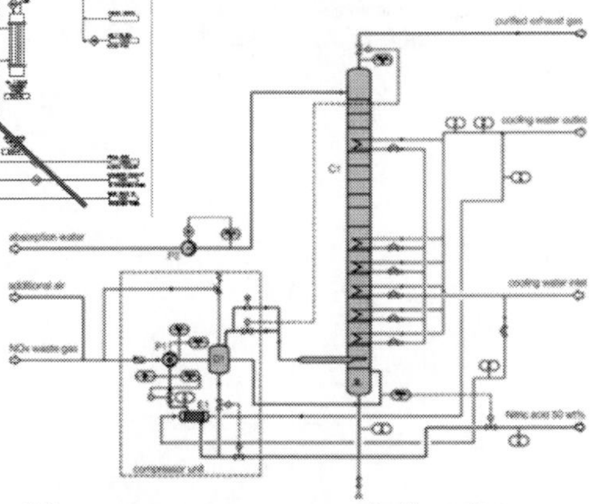

Improved understanding of chemistry

Operation at alternative conditions

Development of specialized internals

➤ Leverage external expertise with proven commercialization experience to obtain readily available off the shelf improvement

* Schematic courtesy of De Dietrich Process Systems, Inc.

LEVEL 3 (NEXT GEN) EXAMPLES #2 & #3

COMBINING MULTIPLE OPERATIONS INTO ONE INTEGRATED UNIT

DIVIDED WALL DISTILLATION COLUMN

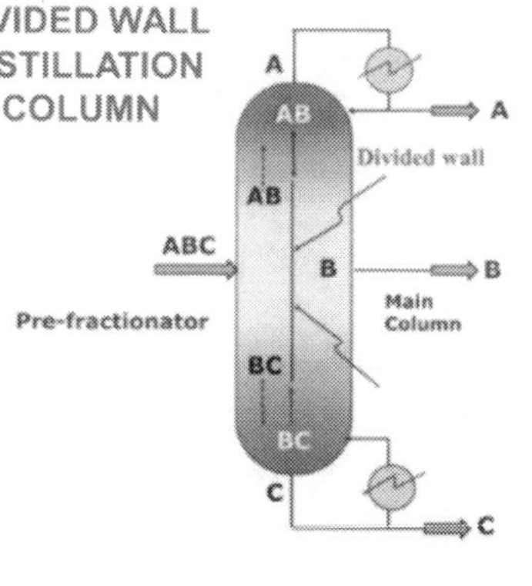

MULTIPLE EVAPORATORS AND STRIPPERS IN ONE UNIT*

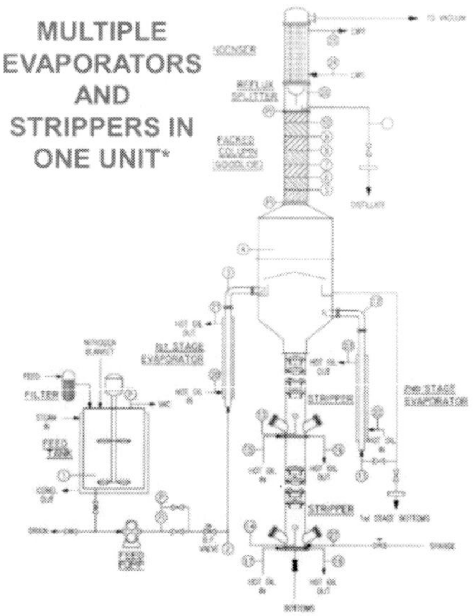

Advanced intensification will require firm foundational information, use of external expertise, a thorough pilot, and a multiple day "semiworks style" validation.

* Schematic courtesy of Artisan Industries Inc.

No. 12

LEVEL 3 (NEXT GEN) ADD'L EXAMPLES

- BATCH TO CONTINUOUS
- IN-LINE OR ON-LINE ANALYTICAL
- ADVANCED REACTOR DESIGNS
- MULTI-STAGE CENTRIFUGAL CONTACTORS
- MULTI-FUNCTIONAL EXTRUDERS

No. 13

BENEFITS

- Capital cost reduction of 20+% per kilogram for assets
- Operating cost reduction of 5 to 10% per kilogram
- Establish firm foundation for low/no-cost investment to expand existing assets

No. 14

THANK YOU FOR YOUR ATTENTION!!!

ANY QUESTIONS?

No. 15

Risk and Uncertainty Analysis of Error Propagation in Process Design and Simulation—Approaches and Modeling Results

Vasquez, V. R., University of Nevada, Reno

Originally presented on: 6/8/2016 14:00:00 - 14:30:00

Risk and uncertainty analysis of error propagation in process design and simulation play an important role in the determination of error bounds of model predictions, design of safety factors, and risk estimation among other. Chemical process design and simulation rely heavily on computer-aided methodologies and thermophysical properties. The latter are commonly estimated through the use of empirical models, correlations, and theoretical models whose parameters are usually estimated from experimental data. In this presentation, we discuss how the issues of random and systematic errors in thermodynamic models affect their predictions and the impact of these on design and simulation with the analysis of a variety of situations and examples. Using Monte Carlo-based methods, we show that there is significant uncertainty in the design and operations of chemical processes when one considers the seemingly small uncertainty in data for vapor-liquid and liquid-liquid equilibria. A common approach in the chemical processing industries is to apply safety factors in design to account for the effects of uncertainties. However, Monte Carlo uncertainty analysis can be used to quantify the uncertainty and lead to rationalization of the safety factor approach. For example, sensitivity analysis of binary interaction parameters typical of models such as NRTL and UNIQUAC show that small variations in these translate into significant differences in performance evaluation of unit operations such as distillation and liquid-liquid extraction. The development of thermodynamic models from experimental data requires the regression of model parameters, which we use as input to the Monte Carlo simulations. However, the classical application of maximum likelihood methods for regression assumes that the model is inherently more accurate than the data. For phase equilibrium data, often the inverse is true. We show, with examples, the importance of separating the effects of different types of uncertainty such as those caused by model uncertainty and those by experimental uncertainty. Monte Carlo methods show to be a valuable tool in developing strategies to make this separation. If time allows, we will briefly discuss current initiatives in our research group towards developing methods for real-time risk assessment of patient safety in clinical processes using machine learning methods and big data; ideas that can be easily extended to many operations of the CPI.

Quality Risk Assessment in Late Phase Drug Substance Process Development

Iyer, V., Bristol-Myers Squibb Company

Originally presented on: 6/8/2016 14:30:00 - 15:00:00

This presentation outlines the process and tools that Bristol-Myers Squibb uses to conduct Quality Risk Assessment (QRA), to achieve consistent drug substance quality and production process performance. The QRA is carried out at multiple stages during the drug substance development lifecycle, with the knowledge requirements and risk mitigation plans increasing with each subsequent iteration. During this process, the quality attribute(s) (QAs) of the drug substance are established, and risk based linkages from the process raw materials and production procedures to these QAs are established. From this analysis, high risk items are identified to focus further development work, improve scientific understanding of the process and/or develop additional process controls. A Failure Modes and Effects Analysis (FMEA) methodology is employed that identifies failure modes (for the QAs) and potential causes (variability in process parameters, chemical equivalents, etc.). The impact of the failure mode (severity) is determined, together with the probability and detection of its occurrence. Understanding failure modes leads to the development of a control strategy designed to ensure the proposed acceptance criteria are consistently met. Ultimately, this risk assessment process establishes a thorough understanding of the process variables and their collective impact to the drug substance QA, thereby effectively managing risk to quality.

PDS Process Development Symposium

DSM
BRIGHT SCIENCE. BRIGHTER LIVING.

Efficient Biocatalytic Processes Developed for Chiral and Functional Intermediates Production

AIChE
Process Development Symposium

Iwona Kaluzna
DSM Innovative Synthesis

Providence,
June 9th, 2016

HEALTH · NUTRITION · MATERIALS

Outline

- Introduction to Biocatalysis
- Application of Cytochrome P450 Technology
- Application of DERA Technology
- Application of Hydratase Technology

HEALTH · NUTRITION · MATERIALS

DSM
BRIGHT SCIENCE. BRIGHTER LIVING.

From Chemical to Enzymatic Synthesis

Biocatalysis is the use of enzymes as they are provided by nature and improved in the laboratory as catalysts for organic chemistry

Why Would Industry Utilize Biocatalysis?

- Chemo-selectivity, regio-selectivity, enantioselectivity, less unwanted by-products
- No (little) protective groups, reduction of process steps, multi step catalysis (no isolation of intermediates)
- Water based chemistry, mild conditions (ambient temperature, pH and pressure), mild activating agents
- Catalyst recycle possible, catalyst improvement by directed evolution (higher activity, better selectivity, better stability)

Less environmental impact, better product quality & cost reduction

DSM
BRIGHT SCIENCE. BRIGHTER LIVING.

Biocatalysis in Route Scouting

- Biocatalysis is typically embedded in a multi-step chemical process
- Typically not just replacing a chemical reaction for a biocatalytic step
- Enabling novel, shorter routes towards the target product
- From alternative raw materials

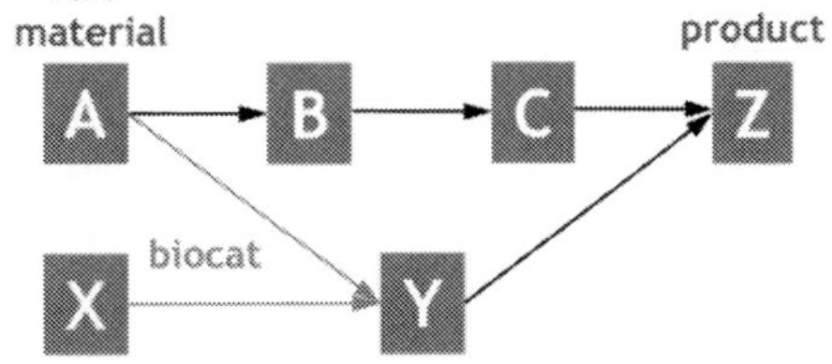

DSM
BRIGHT SCIENCE. BRIGHTER LIVING.

Page

Cytochrome P450 Technology

HEALTH · NUTRITION · MATERIALS

DSM
BRIGHT SCIENCE. BRIGHTER LIVING.

CYP450s Catalyze a Broad Reaction Range

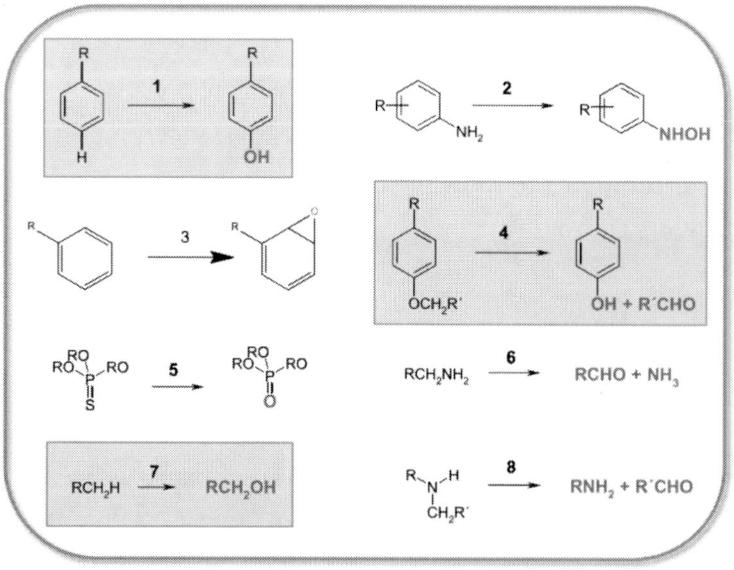

1 aromatic hydroxylation
2 arom. aminohydrox.
3 aromatic epoxidation
4 *O*-dealkylation
5 desulfurylation
6 deamination
7 aliphatic hydroxylation
8 *N*-dealkylation

Introduction of oxygen into non-activated C-H bonds

DSM
BRIGHT SCIENCE. BRIGHTER LIVING.

The P450 Technology at DSM

Reported until nowadays in the literature

- For industrial fine chemical
space time yield → 0.1 g/L/h and a product concentration of 1 g/L

- For pharmaceuticals and metabolites
space time yield → 0.001 g /L/h and a product concentration 0.1 g/L

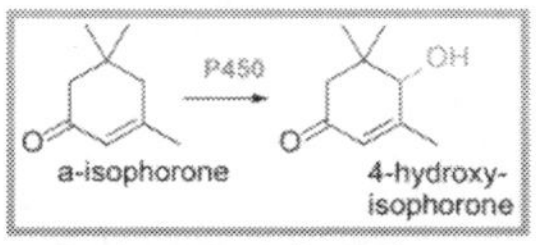

Productivity 1.5 g/L/h
Product conc 10 g/L

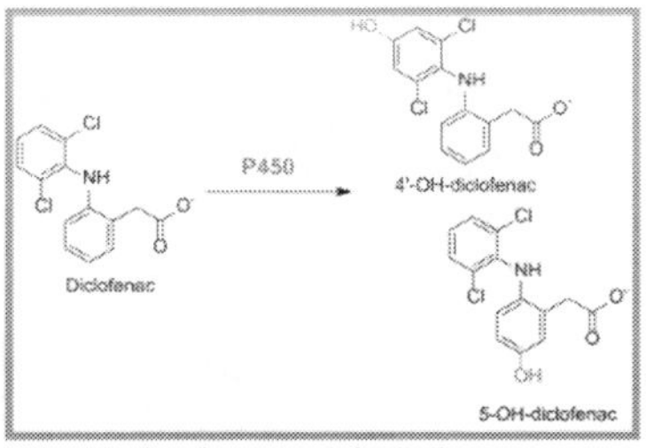

Productivity 1.0 g/L/h
Product conc 4 g/l

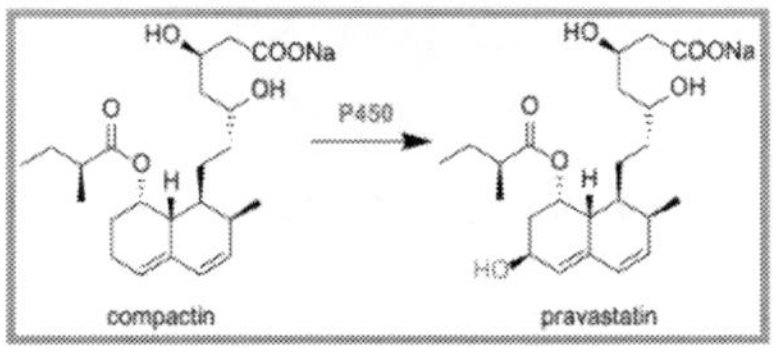

Fermentation titers >2g/kg

DSM
BRIGHT SCIENCE. BRIGHTER LIVING.

Diversity: The P450 collection

Wild type and engineered P450s

Wild type

Various prokaryotic & eukaryotic	~150

Engineered

Bacillus library	7 000
Amycolatopsis library	5 000

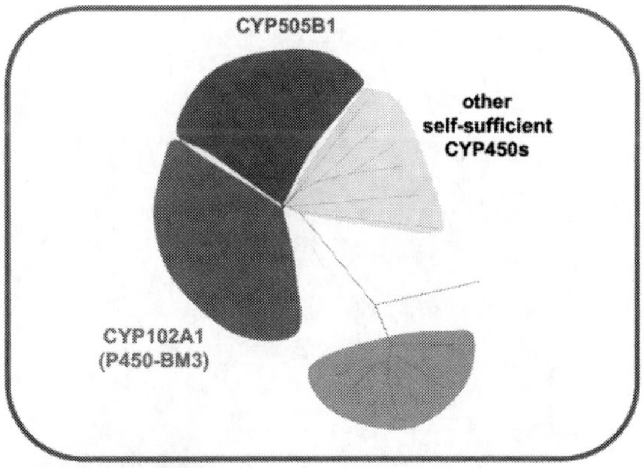

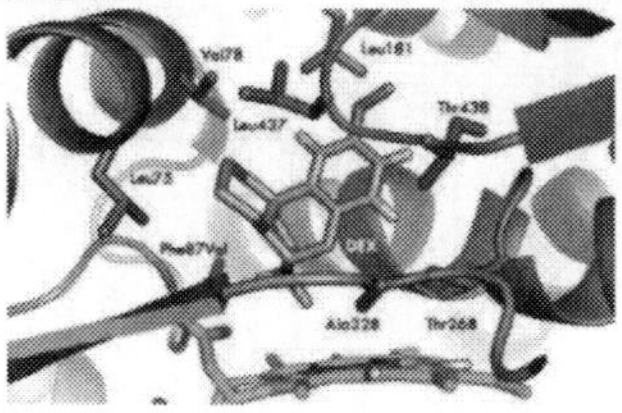

DSM
BRIGHT SCIENCE. BRIGHTER LIVING.

A platform for Exploring Substrate/Product Space of P450s

*flow*NMR	UPLC-MSMS
	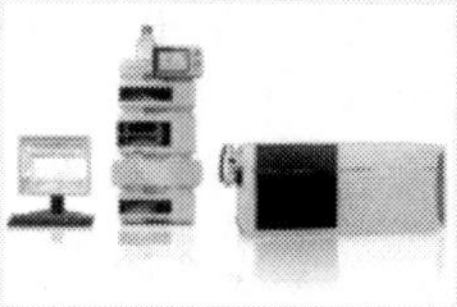
• Fast - 1' / sample • Accurate - even without reference • Broadly applicable - for almost all enzymatic reactions	• Fast - 4'/ sample • Accurate & Sensitive - few ng/ml • Reliable - analysis of complex mixtures possible

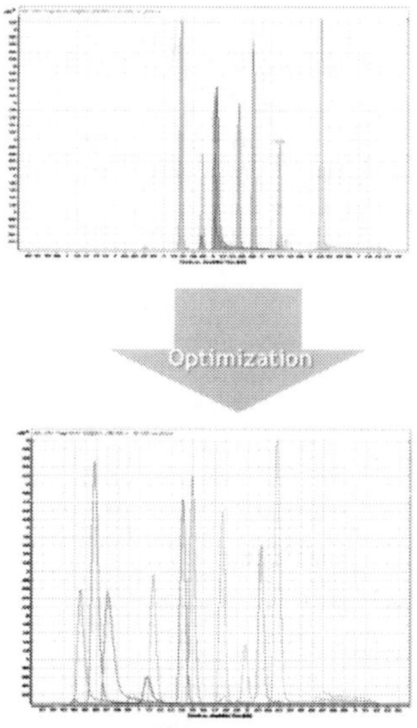

Total run time ~ 3.6 min using an UPLC column (injection to injection)

7 000 variants characterized towards 13 substrates within 17-20 days

DSM
BRIGHT SCIENCE. BRIGHTER LIVING.

Enzymatic Process Considerations -Scale up

Parameters optimized in house

P450 Process

- ✓ Co-factor regeneration
- ✓ Redox system
- ✓ H_2O_2 formation
- ✓ Oxygen supply
- ✓ Low enzyme stability

- ✓ Temperature control
- ✓ pH control
- ✓ Mixing (shear forces)
- ✓ Substrate supply
- ✓ Down-stream processing (e.g. enzyme removal, product recovery)

DSM
BRIGHT SCIENCE. BRIGHTER LIVING.

Biohydroxylation *vs* Chemical Oxidation

- Unique capability of P450s to perform allylic oxidations
- Chemical oxidation has to occur from ß-isophorone
- Conversion of α- to ß-isophorone is low yielding and energy intensive
- Cytochrome P450 process on 100 L scale
- Production of > 1kg 4-hydroxy-alpha-isophorone with productivity of 1.5 g/L/h and substrate load of 10 g/L
- Obtained productivity 150 higher compared to existing processes

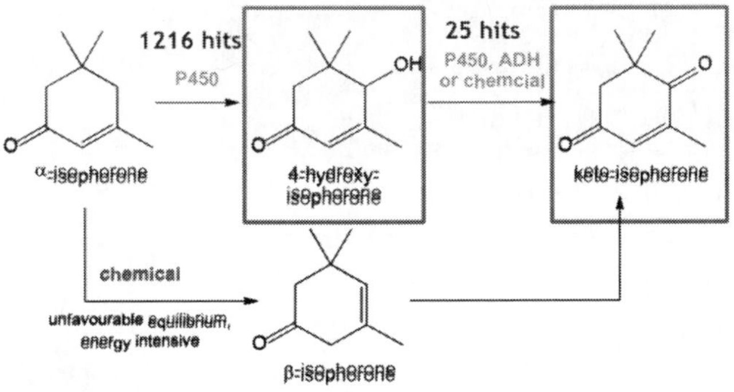

DSM
BRIGHT SCIENCE. BRIGHTER LIVING.

Systematic Optimization

Application:

used for parameter optimization

•pH, temperature, protein amounts

•Substrate concentration

Advantage:

•pH control, temperature control

•Improved mass balance (condenser)

•3 reactions in parallel

•Sampling, reaction followed in time

Drawback:

•Limited control of O_2 input

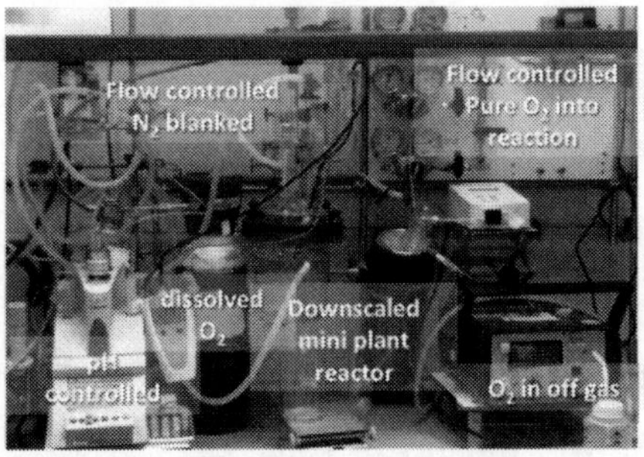

Application:

•Investigation of product isolation

•Gram scale sample product

Advantage:

•Direct up-scale from 20 ml system

•Air or pure oxygen used

•Better mass balance

•O_2 sensors enable O_2 mass balance

DSM
BRIGHT SCIENCE. BRIGHTER LIVING.

Foam Formation by Denatured Protein

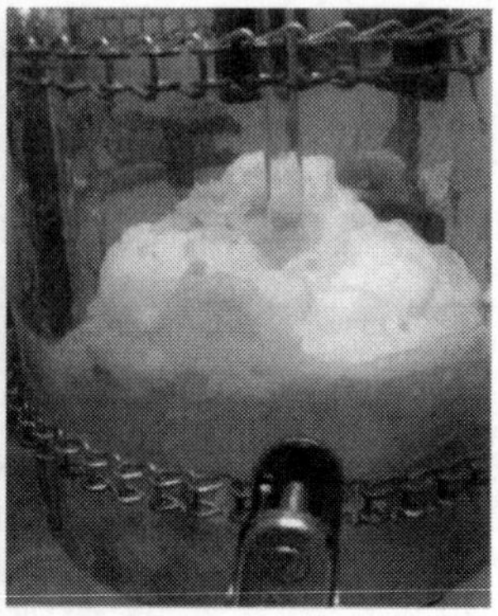

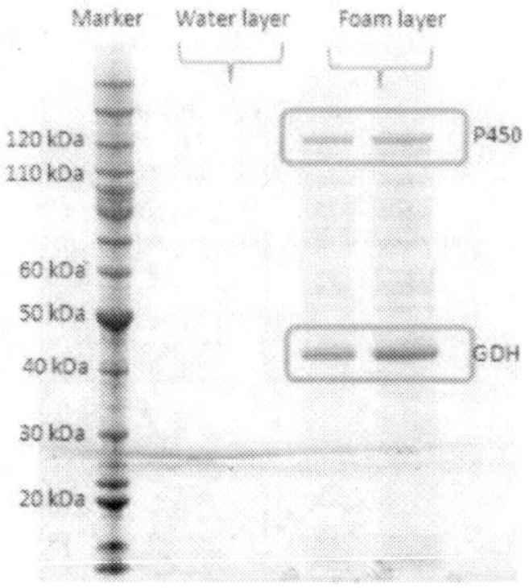

- > 80% of total protein in foam layer
- No active enzyme available for target reaction

→ Foam needs to be avoided

→ Slower aeration

DSM
BRIGHT SCIENCE. BRIGHTER LIVING.

Slower Aeration to Prevent Foam Formation

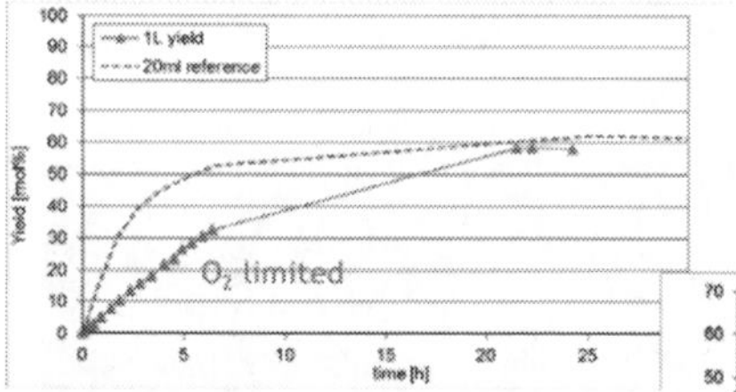

Application of pure oxygen

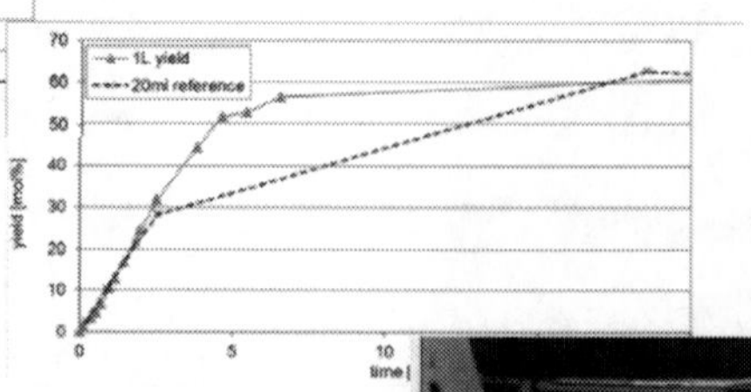

- Still some foam formation
- Slower but linear reaction progress

→ O_2 limited

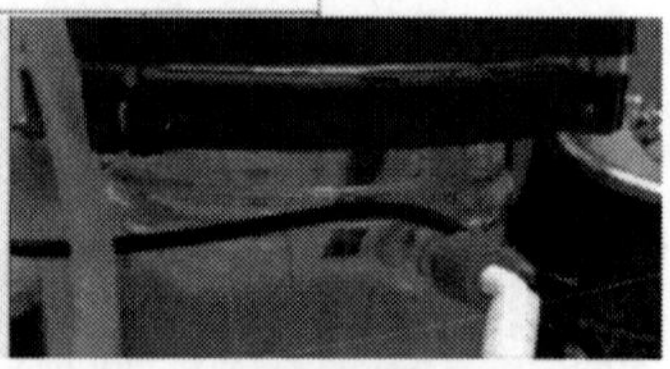

- No foam formation
- No O_2 limitation
- Identical behavior on 20 mL and 1 L scale

DSM
BRIGHT SCIENCE. BRIGHTER LIVING.

Mini Plant Run

100 L scale

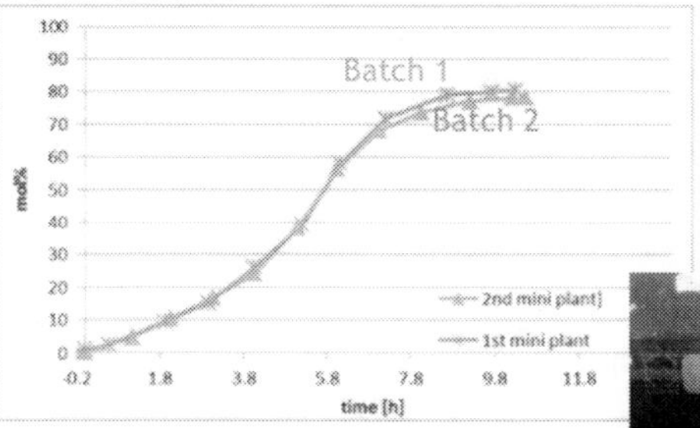

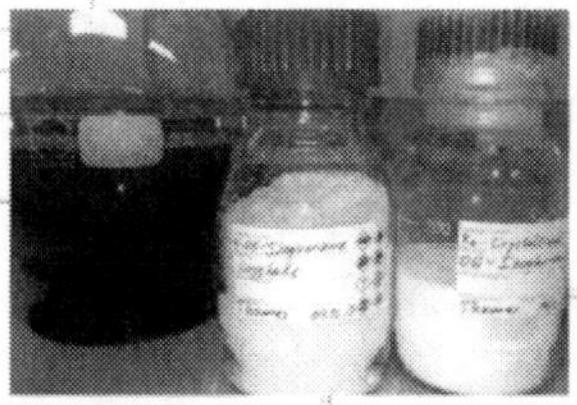

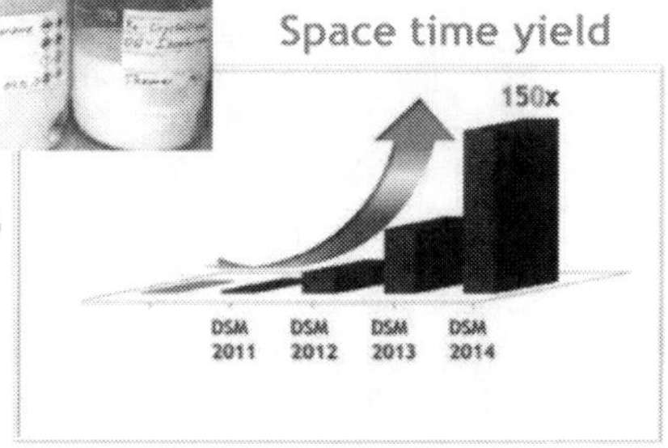

- Almost 100% identical behavior of both batches
- >1kg of product isolated
- 100 L slower compared to 1 L reference

DSM BRIGHT SCIENCE. BRIGHTER LIVING.

Downstream Processing of Aqueous Reaction

- Addition of 50 vol.-% methanol @55°C overnight
- Addition of dicalite as body feed
- Filtration over Seitz filter to precipitate enzyme/cells

DSM BRIGHT SCIENCE. BRIGHTER LIVING.

Downstream Processing: Extraction

- Evaporation of methanol
- Extraction with ethyl acetate (volume ratio: ~ 1:1)

Almost 100% extraction yield after one step

- Rotating evaporator

➡ Product yield Batch 1: ~ 88%

➡ Product yield Batch 2: ~ 86%

DSM
BRIGHT SCIENCE. BRIGHTER LIVING.

DERA Technology

HEALTH · NUTRITION · MATERIALS

DSM
BRIGHT SCIENCE. BRIGHTER LIVING.

Lowest cost process to Statins

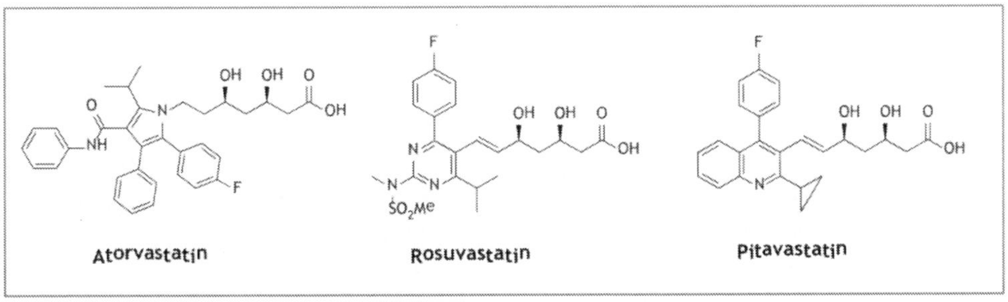

- Statins are the 1st choice in treating dyslipidemia (high cholesterol levels)
- Leading to cardiovascular diseases, the global no. 1 lifestyle disease
- According to WHO 49% of world health population suffers from high cholesterol levels

SINOCHEM | DSM
DSM Sinochem Pharmaceuticals

Retrosynthesis

Chiral Intermediate and Connection of Both Parts

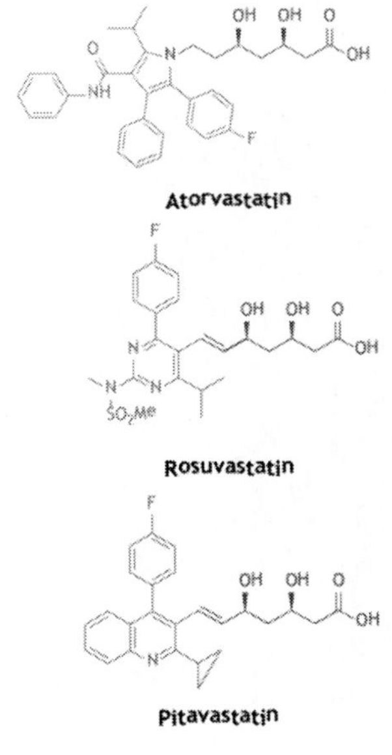

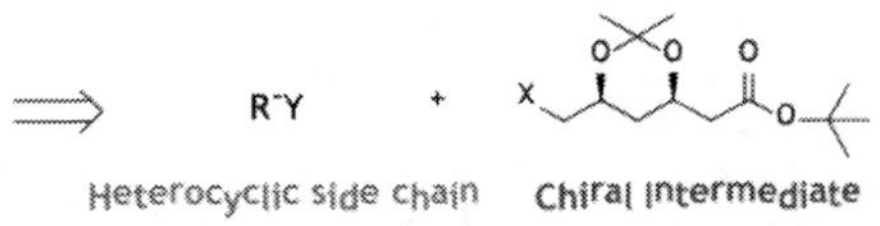

Review: *Historic Overview and Recent Advances in Synthesis of Super-Statins*, Zdenko C. *Current Organic Chemistry* 2010, *14*, 816-845.

SINOCHEM | DSM
DSM Sinochem Pharmaceuticals

Chiral Intermediate by DERA

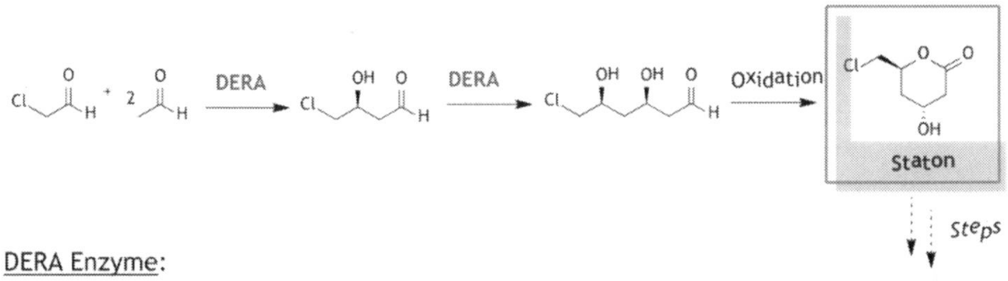

Atorvastatin
Rosuvastatin
Pitavastatin

DERA Enzyme:

- Both chiral centers fixed in a single step!
- Chain elongation not required: all carbon atoms in place
- Simple and readily available starting materials
- One common intermediate Staton for all statins
- In total 3-4 steps to "chiral intermediate"
- Wild type DERA improved by directed evolution targeting productivity and improved tolerance towards chloro acetaldehyde (WO 2003/006656)

DERA = deoxyribose phosphate aldolase
Lit see: Wong *et al.* (1995) *J. Am. Chem. Soc.*, 117, 3333,

SINOCHEM | DSM
DSM Sinochem Pharmaceuticals

SINOCHEM | DSM
DSM Sinochem Pharmaceuticals

Statins - Common intermediate

Proprietary, enzymatically produced intermediate:
Independent of intermediates used by industry

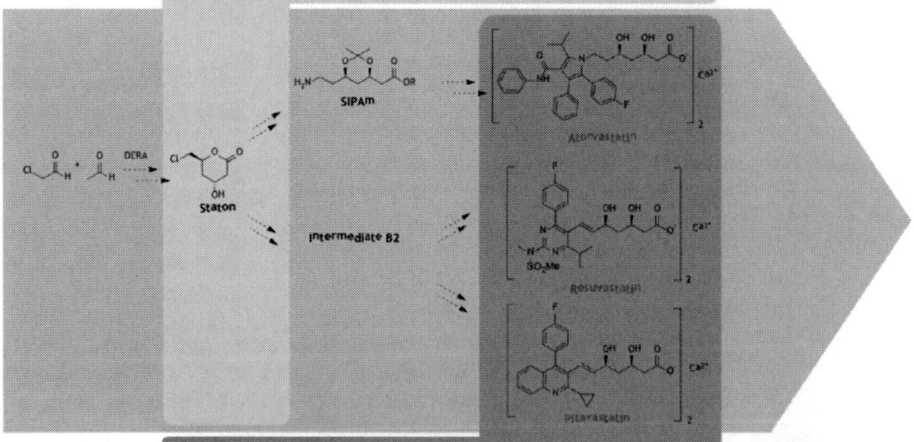

High yields and efficient side chain coupling lead to unrivalled purity and low carbon footprint. Compliant with the strictest (draft) monographs

DSM
BRIGHT SCIENCE. BRIGHTER LIVING.

Rosuvastatin - New Process

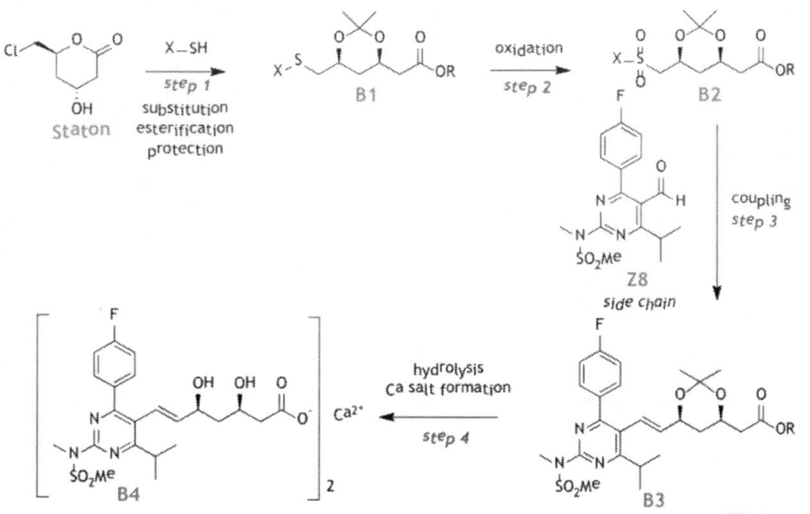

Separation of Biocatalyst from the Reaction

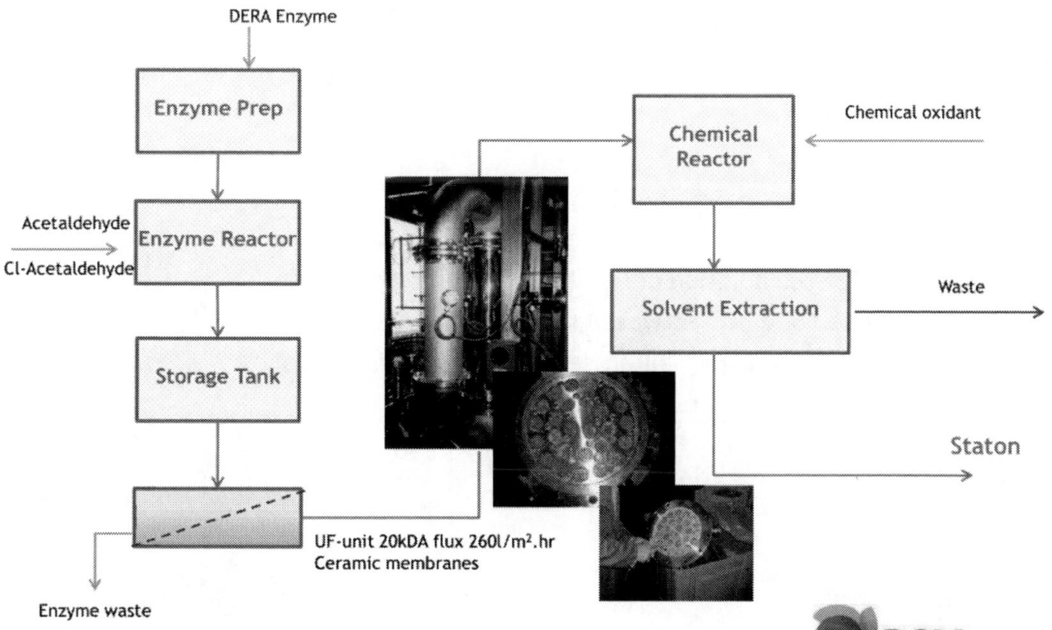

Hydratase Technology

HEALTH · NUTRITION · MATERIALS

DSM
BRIGHT SCIENCE. BRIGHTER LIVING.

Hydratase Technology

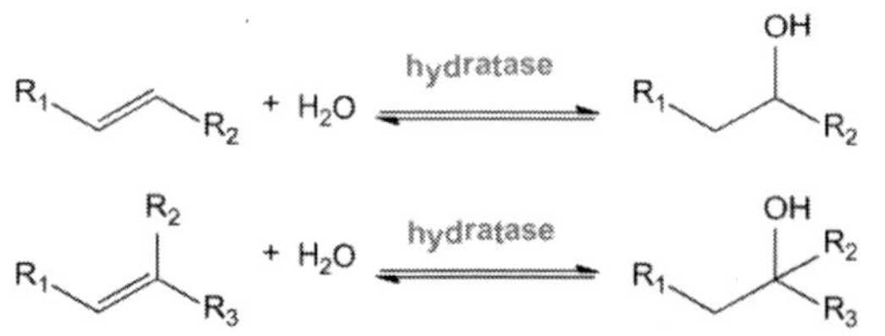

- ➢ Catalyze the addition of H_2O to C=C double bonds
- ➢ Activated and non-activated substrates
- ➢ Producing chiral or achiral, secondary or tertiary alcohols
- ➢ Generally very regio- and enantioselective

DSM
BRIGHT SCIENCE. BRIGHTER LIVING.

Hydratase Platform

10-Hydroxy-Stearic Acid (10-HSA) by Oleate Hydratases

- A large collection of different hydratases are available in house
- Application of the hydratase technology resulted in efficient process for 10-HSA production on ton scale
- First commercial grade product, 10-HSA will be introduced to the market in Q2 2017

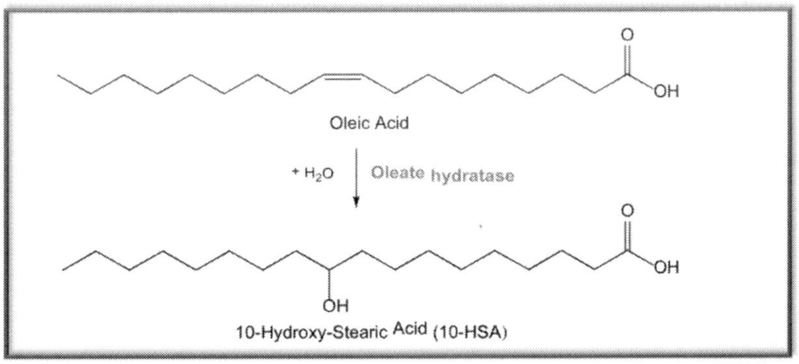

DSM
BRIGHT SCIENCE. BRIGHTER LIVING.

Enzymatic Process Considerations -Scale up

Parameters optimized in house

Hydratase Process

- ✓ Temperature
- ✓ pH control
- ✓ Mixing (shear forces; mass transfer limitation)
- ✓ Substrate supply
- ✓ Downstream processing:
  - Enzyme removal
  - Evaporation
  - Extraction (choosing the right solvent)
  - Product recovery

DSM
BRIGHT SCIENCE. BRIGHTER LIVING.

Mini Plant Run

The reaction was run in the 200 L reactor (100 L reaction volume).

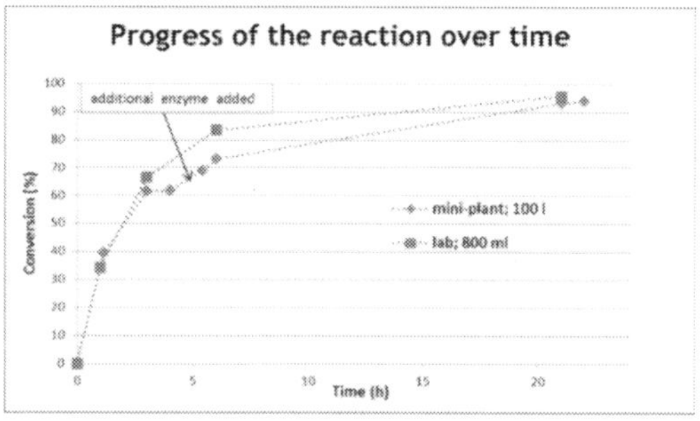

Enzyme reaction

⇩

Enzyme filtration

⇩

Extractions

⇩

Crystallization

⇩

Product filtration

- 10 kg Oleic acid was converted
- Deactivation of the enzyme caused by the high tip speed
- After addition of fresh enzyme and lowering the stirrer speed, the reaction reached 94% conversion.

DSM
BRIGHT SCIENCE. BRIGHTER LIVING.

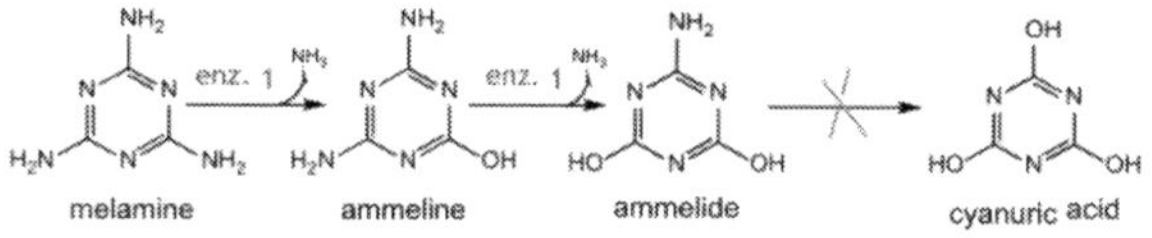

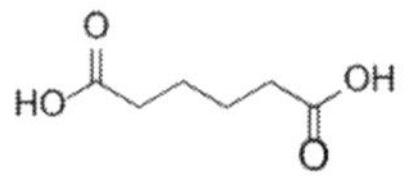

Biobased adipic acid

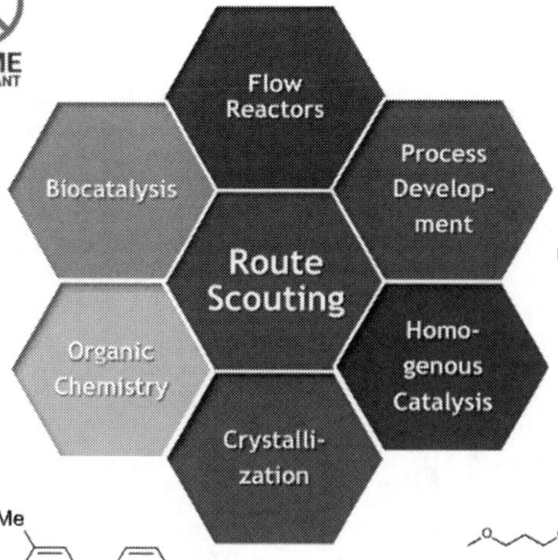

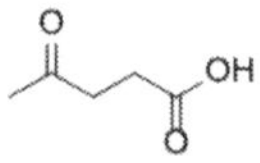

Biobased levulinic acid

3D printed flow reactors

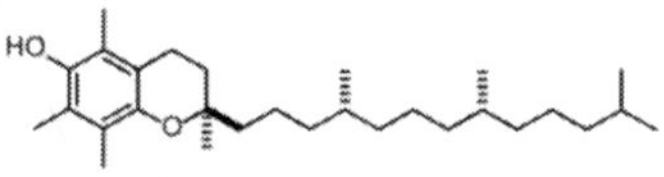

Vitamines

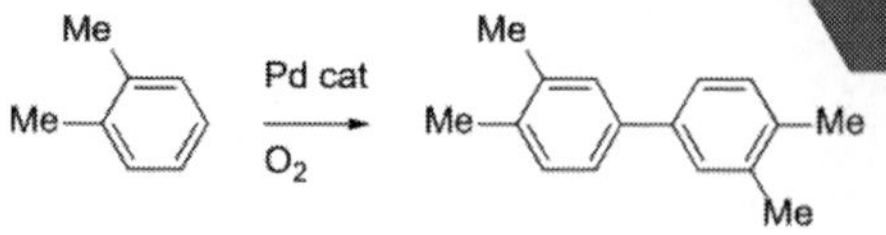

Biaryls

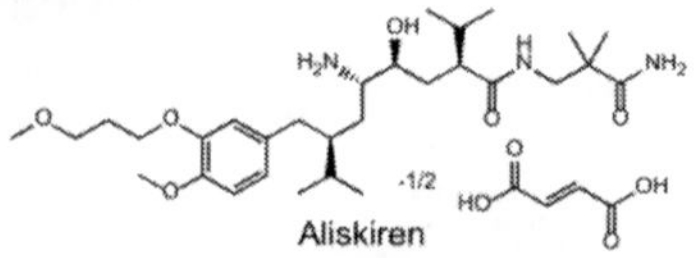

Aliskiren

DSM
BRIGHT SCIENCE. BRIGHTER LIVING.

Contact
iwona.kaluzna@dsm.com
+31 46 476 7058

HEALTH · NUTRITION · MATERIALS

DSM
BRIGHT SCIENCE. BRIGHTER LIVING.

CHARACTERIZING IMPELLER PERFORMANCE WITH EXAMPLES OF PROCESS RESULTS

Richard K. Grenville
Director of Mixing Technology
Philadelphia Mixing Solutions Ltd.

AGENDA

- Impeller geometries.
- Characterizing performance:
  →Power, Flow, Shear & Efficiency.
- Process results:
  - Dispersion of immiscible liquids.
  - Timescales of mixing.
  - Flocculation.
- Conclusions.

HYDROFOIL IMPELLERS

Low-solidity

Po ≈ 0.30

Medium-solidity

Po ≈ 0.70

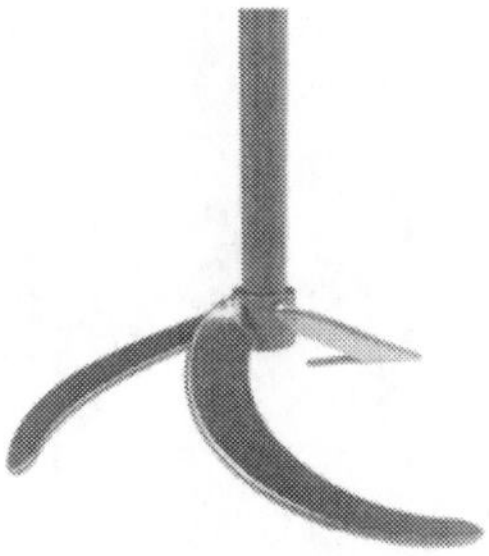

Anti-ragging

Po ≈ 0.30

“Low shear” impellers (Ducoste et al, AIChEJ, 1997).

BLADE TURBINES

Pitched

Po ≈ 1.30

Vertical / Flat

Po ≈ 3.00

DISK TURBINES

Rushton	Smith	PMSL GDX
Po ≈ 5.00	Po ≈ 3.00	Po ≈ 1.00

"High shear" impellers (Ducoste et al, AIChEJ, 1997).

IMPELLER CHARACTERISTICS

- Power number: $Po = \frac{P}{\rho N^3 D^5}$
- Flow number: $Fl = \frac{Q}{ND^3}$
- Diameter ratio: $\frac{D}{T}$
- Projected blade height: $y = \frac{D}{w_P}$

PHILADELPHIA MIXING SOLUTIONS MIXING SOLUTIONS LIMITED

EFFICIENCY

- Mass of Fluid Pumped per unit of Energy Input (Fořt et al., 2010, Chem. Proc. Eng.).

$$\eta = \frac{Fl\rho ND^3}{Po\rho N^3D^3} = \frac{Fl}{PoN^2D^2} \quad kg / J$$

- At equal Power Input per Mass:

$$\eta = 1.18\frac{Fl}{Po^{1/3}}(\frac{D}{T})^{4/3}(\bar{\varepsilon}T)^{-2/3}$$

FLOW EFFICIENCY

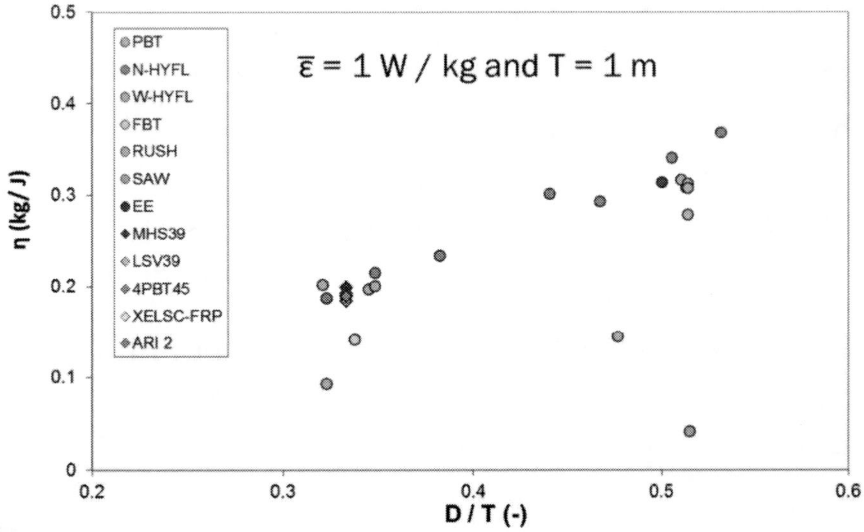

Brown, 2010,MIXING XXII and PMSL

TIME-AVERAGED SHEAR RATE

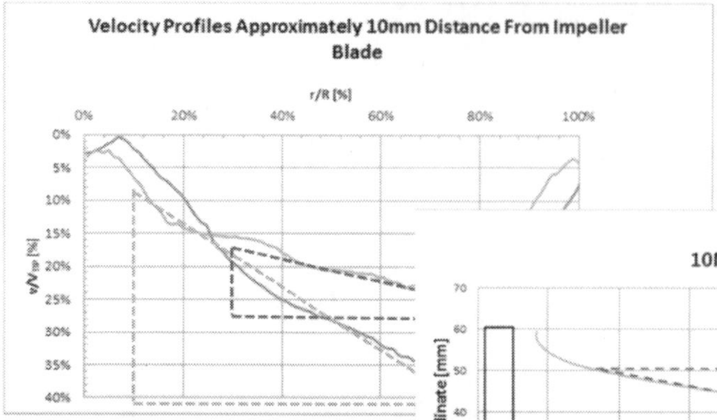

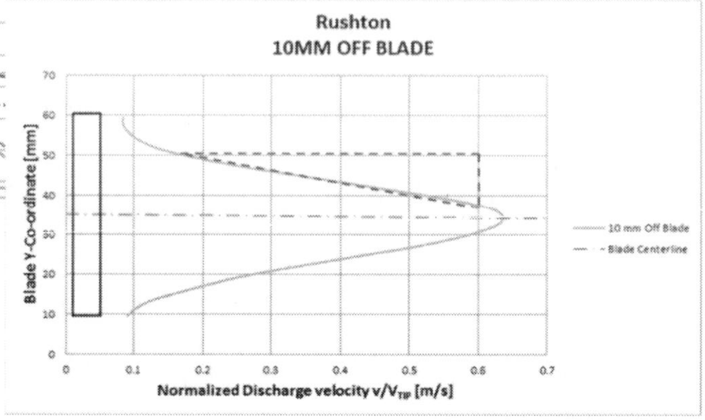

After Oldshue, Fluid Mixing Technology, McGraw-Hill, 1983.

PHILADELPHIA MIXING SOLUTIONS MIXING SOLUTIONS LIMITED

SUMMARY

- Hydrofoils are:
  - Most efficient – hydraulically.
  - Generate lowest time-averaged shear rate.
- Rushton turbines are:
  - Least efficient – hydraulically.
  - Generate highest time-averaged shear rate.

- Drop size measurements in dispersion of immiscible liquids:
  - Process dependent on shear.

2016 Process Development Symposium | Providence, RI| June 7-9, 2016

PHILADELPHIA MIXING SOLUTIONS MIXING SOLUTIONS LIMITED

d_{32} vs. $\bar{\varepsilon}$ – All D = ~ T / 2

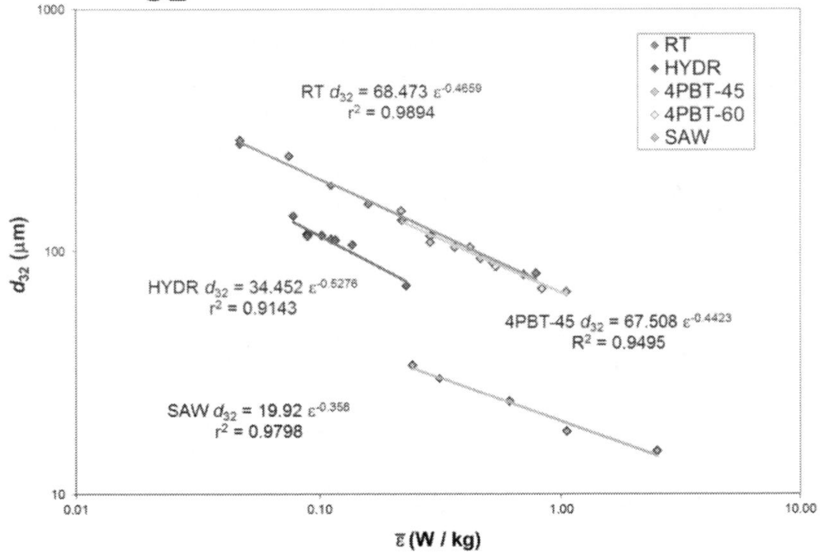

See also Pacek et al., 1999, Chem. Eng. Sci.

TRAILING VORTEX

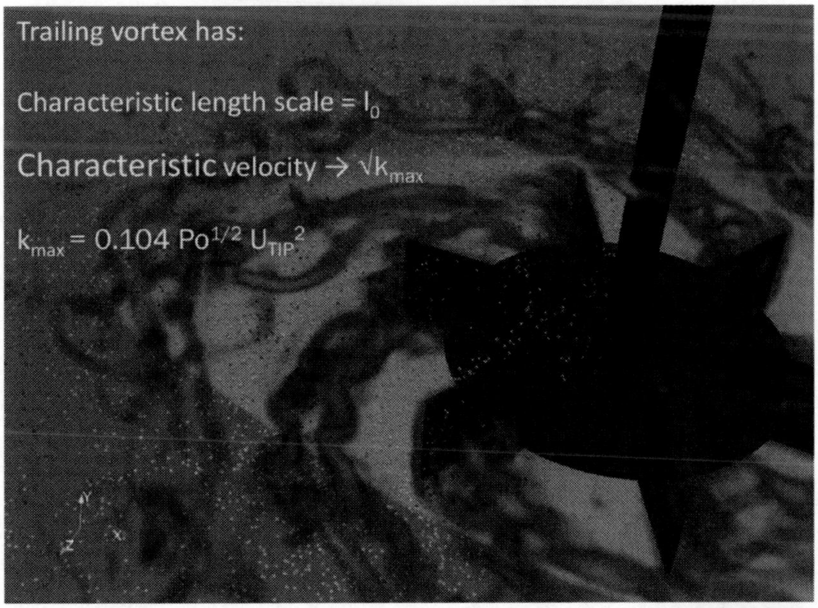

k_{MAX} and ε_{MAX}

- Kinetic Energy and Dissipation Rate within Trailing Vortex (Grenville et al., 2008, AIChE Annual Meeting):

$$k_{MAX} = 0.104 Po^{1/2} U_{TIP}^2 \quad (\pm 10\,\%) \qquad \text{and:} \qquad l_0 = \frac{D}{x}$$

$$\varepsilon_{MAX} = A \frac{k_{MAX}^{3/2}}{l_0} = Ax \frac{(0.104 Po^{1/2} U_{TIP}^2)^{3/2}}{D}$$

$$\varepsilon_{MAX} = 1.04 x Po^{3/4} N^3 D^2$$

2016 Process Development Symposium | Providence, RI| June 7-9, 2016

PHILADELPHIA MIXING SOLUTIONS MIXING SOLUTIONS LIMITED

d_{32} versus ε_{max} – D = ~ T/2

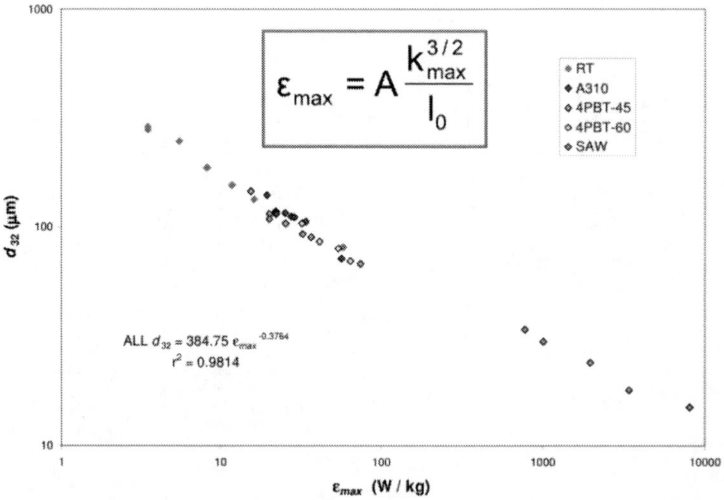

2016 Process Development Symposium | Providence, RI| June 7-9, 2016

PHILADELPHIA MIXING SOLUTIONS MIXING SOLUTIONS LIMITED

MIXING & REACTIONS

- Many reactions are simple:
  - Yield / selectivity unaffected by mixing.

$$A + B \rightarrow P$$

- Many reactions are competitive:
  - Yield / selectivity may be affected by mixing.
  - Kinetic or mixing control?

$$A + B \rightarrow P$$

$$P + B \rightarrow W$$

TEST REACTION

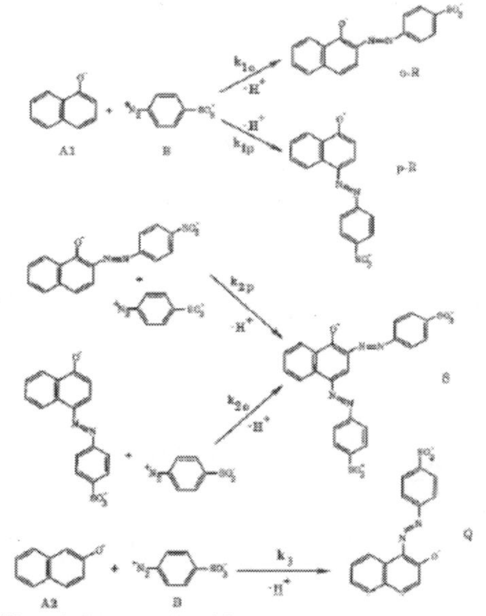

Figure 1. Representation of the new test reaction system.

A1 1-naphthol
B diazotized sulphanilic acid
o-R ortho monoazo dyestuff
p-R para monoazo dyestuff
S bisazo dyestuff

A2 2-naphthol
Q 2-naphthol monoazo dyestuff

Selectivity / yields determined by rate of mixing <u>at feed point.</u>

Bourne et al., 1992, IEC Res.

MIXING TIMESCALES

- Macromixing → at scale of the vessel:
  - Time for vessel contents to become homogeneous.
  - Blend time.
- Micromixing → at scale of smallest turbulent eddies:
  - Deformation of fluid elements (vortex stretching, engulfment).
  - Diffusion.
- Mesomixing → at scale of feed plume:
  - Turbulent exchange of fresh feed with surrounding fluid.
  - Erosive reduction in scale of "blobs" of feed.

Baldyga & Pohorecki, 1995, Chem. Eng. J.

DAMKOHLER NUMBER

$$Da = \frac{r_{React}}{r_{Mix}} = \frac{t_{Mix}}{t_{React}}$$

$r_{Mix} >> r_{React}$

$Da << 1$

Kinetically controlled.

If chemistry (kinetics) are fixed:

$Da \propto t_{Mix}$

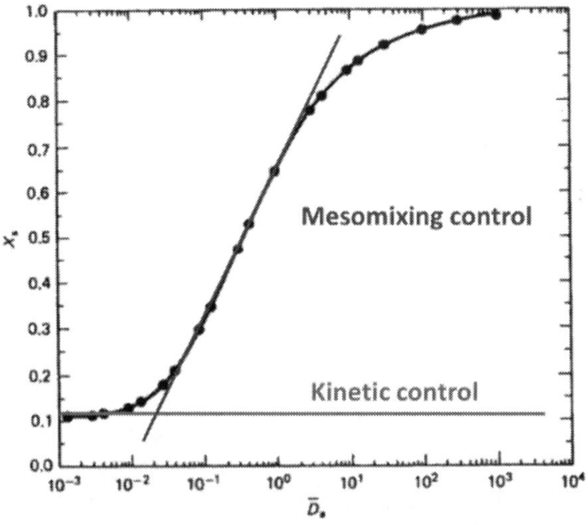

2016 Process Development Symposium | Providence, RI| June 7-9, 2016

EFFECT OF FEED RATE

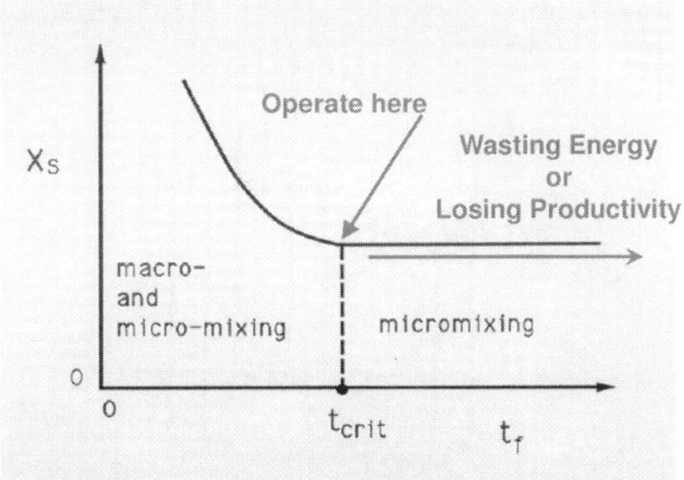

Fig. 1 Product distribution (Xs) as a function of feed time (t_f) for a semi-batch reactor

Baldyga et al, 1993, Chem. Eng. Sci.

PLUME SURFACE AREA

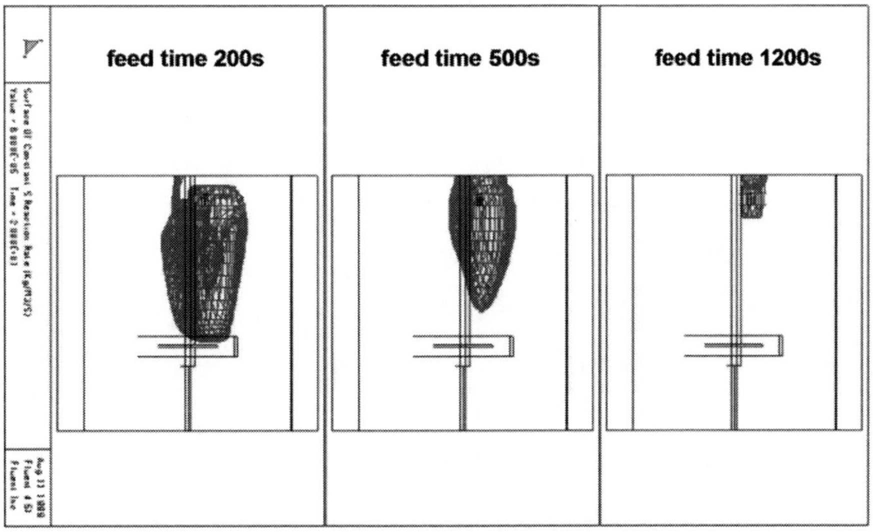

Hannon *et al.*, *MIXING XVII*, 1999

FEED LOCATION

Location B makes 6 -7 x more waste than Locations A or D.

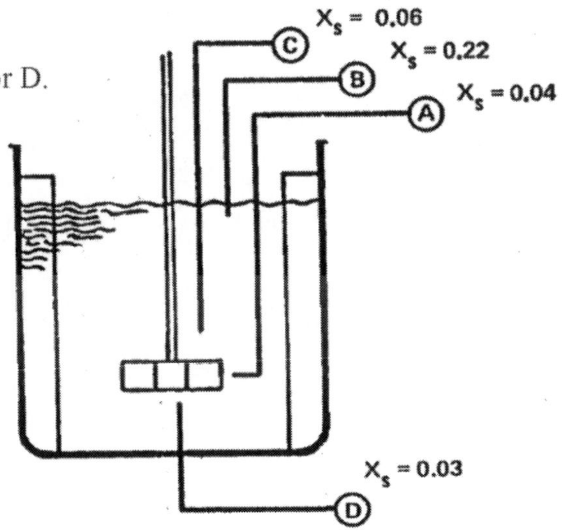

Tosun, 1988, 6th Europ. Conf. on Mixing

EXAMPLE

Acid + Base → Acidic Product (P) + Water

Acidic Product + Base → Inert By-product (W) + Water

- Reaction products are equally soluble:
  - Cannot be easily separated.
- Product sold as solution of P in water.:
  - W is inert in product.
  - Sold on basis of concentration of P.
- Mixture is diluted to make product:
  - Higher P : W ratio → more dilution water added.

PHILADELPHIA MIXING SOLUTIONS MIXING SOLUTIONS LIMITED

LAB STUDIES

- Tests carried out at lab-scale in Hood.
- If mixing is slow product loses "quality".
- Conditions:
  - 1 litre vessel (T = 99 mm) with single Hydrofoil (D = 54 mm).
  - Feed rate, q = 1 and 3 ml / min.
  - Impeller speed = 150 – 600 RPM.
  - Feed at surface and at impeller tip (through dip pipe).

RAW DATA versus SPEED

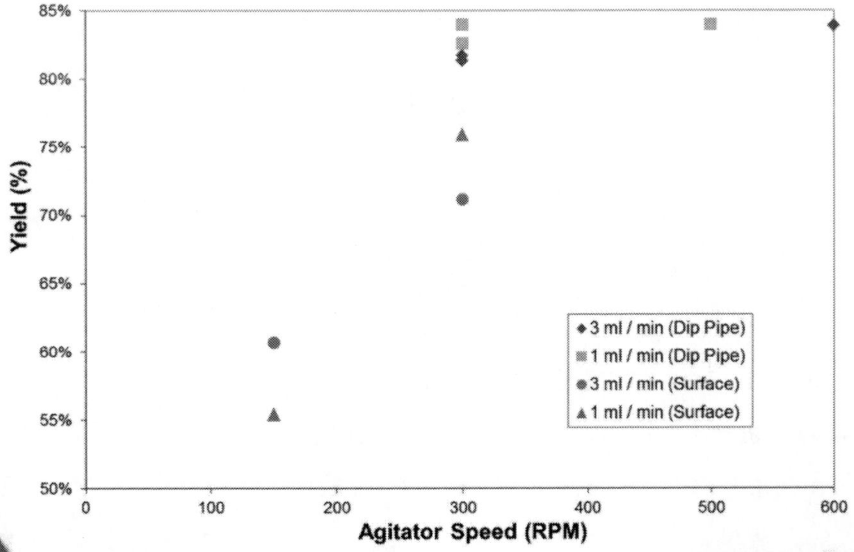

YIELD versus POWER PER MASS

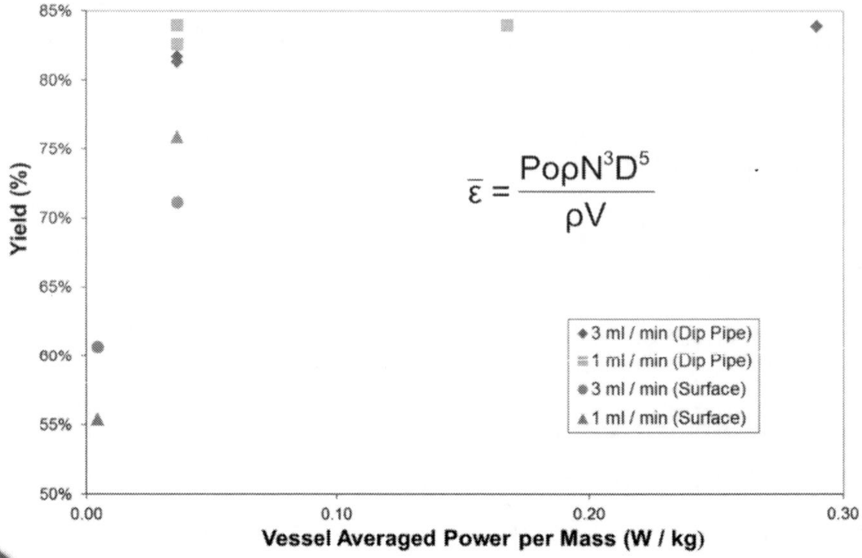

PHILADELPHIA MIXING SOLUTIONS · MIXING SOLUTIONS LIMITED

CALCULATION OF MIXING RATE

- Meso-mixing rate is function of power per mass and mean flow (velocity) at feed location:

$$r_M = \frac{1}{t_M} = 0.46\left(\frac{\overline{U}\varepsilon}{q_B}\right)^{1/3}$$

$$r_M = 0.46\left(\frac{J\overline{U}K\bar{\varepsilon}}{q_B}\right)^{1/3}$$

- Multipliers for impeller feed:
  - At surface: J = 0.25 K = 1.0
  - At impeller: J = 1.0 K = 92

2016 Process Development Symposium | Providence, RI| June 7-9, 2016

PHILADELPHIA MIXING SOLUTIONS · MIXING SOLUTIONS LIMITED

YIELD versus MESOMIXING RATE

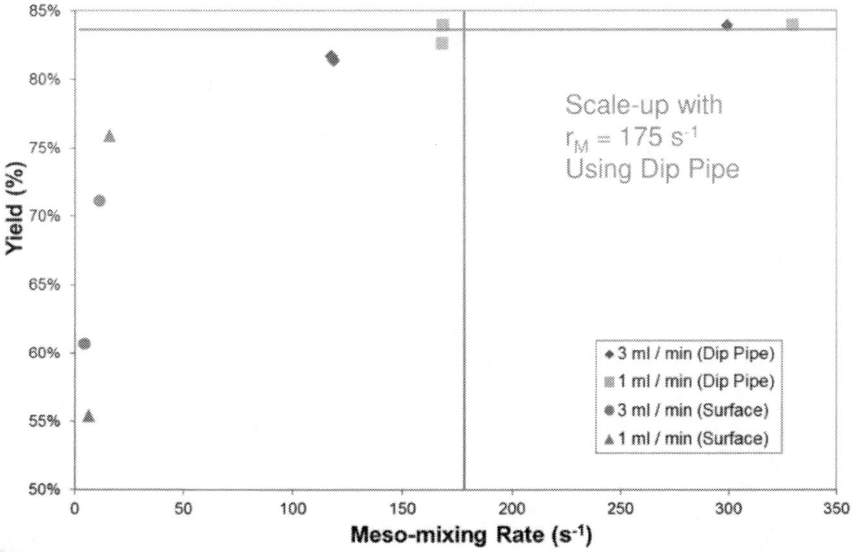

FLOCCULATION

- Removal of fine particles (sub-micron) from liquid:
  - Water and Waste Water Treatment.
  - Polymer processing.
- Flocculation increases particle size:
  - Agglomerates of many particles.
  - Easier to remove by sedimentation or filtration.
- Final floc size determined by equilibrium between growth and breakage.
- Time affects floc shape and density:
  - Breakage and re-flocculation.

MECHANISM OF FLOC GROWTH

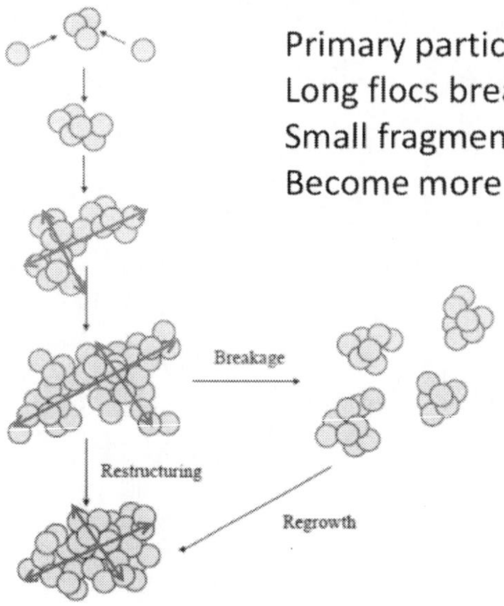

Primary particles collide and agglomerate.
Long flocs break.
Small fragments collide and reorganize.
Become more compact over time.

Spicer, 1997, PhD Thesis, Univ. of Cincinnati

NAMFL vs. TIME

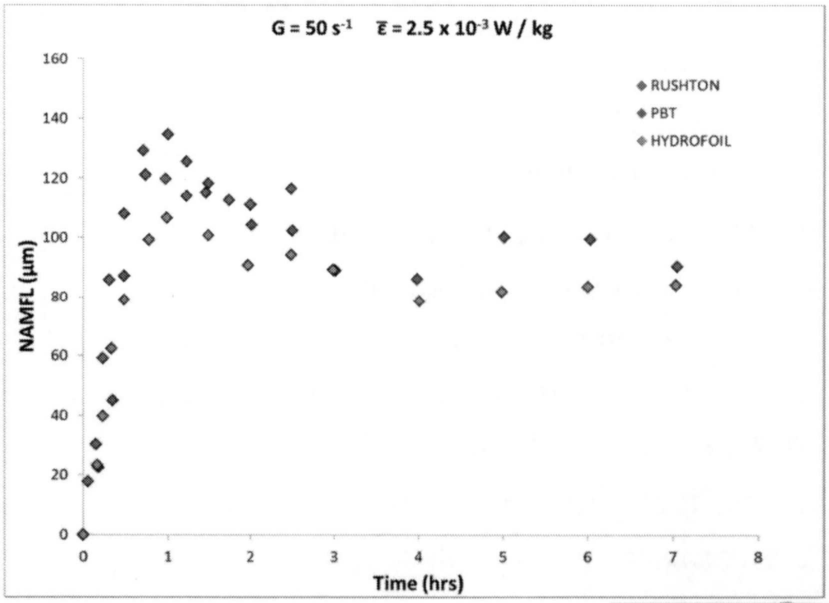

NAMFL vs. TIME – EQUAL ε_{MAX}

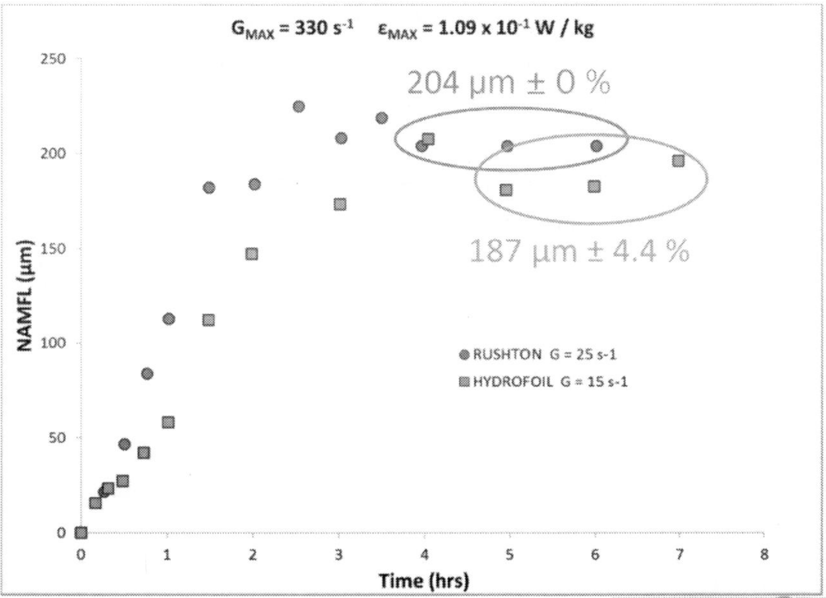

2016 Process Development Symposium | Providence, RI| June 7-9, 2016

PHILADELPHIA MIXING SOLUTIONS
MIXING SOLUTIONS LIMITED

AVG. ≡m. NAMFL vs. ε_{MAX}

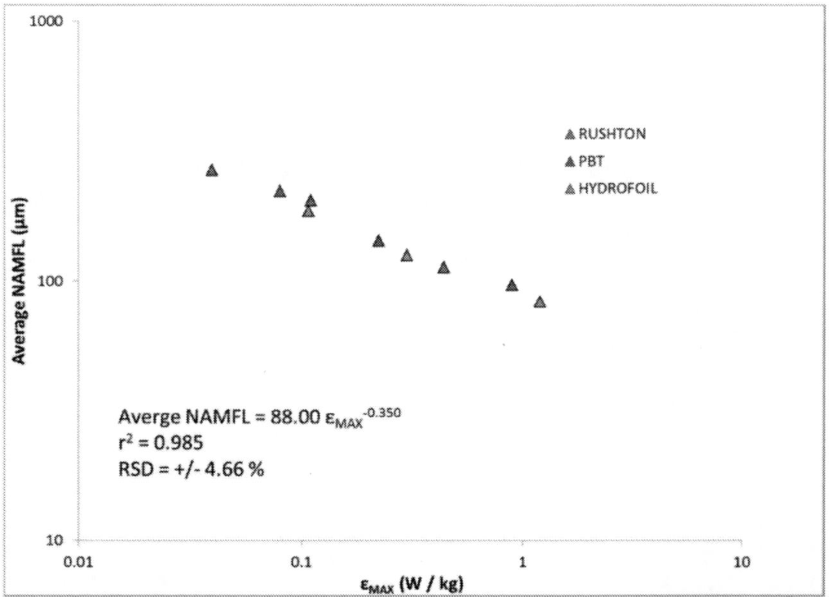

2016 Process Development Symposium | Providence, RI| June 7-9, 2016

PHILADELPHIA MIXING SOLUTIONS
MIXING SOLUTIONS LIMITED

CONCLUSIONS

- Impeller characteristics can be quantified:
  - Flow → Hydraulic Efficiency.
  - Shear → Maximum Energy Dissipation Rate in Trailing Vortex.
- Process result:
  - Global process → Vessel average mixing rate:
    - Blend / circulation time.
  - Local process → Trailing vortex mixing rate:
    - Area generation.
    - Mesomixing.
- This understanding must be applied to scale-up:
  - Identifies "most efficient" impeller type to use.
  - Allows lab or pilot data taken with one impeller to be applied to design of another at larger scale.

BACK-UP CHARTS

NAMFL vs. TIME

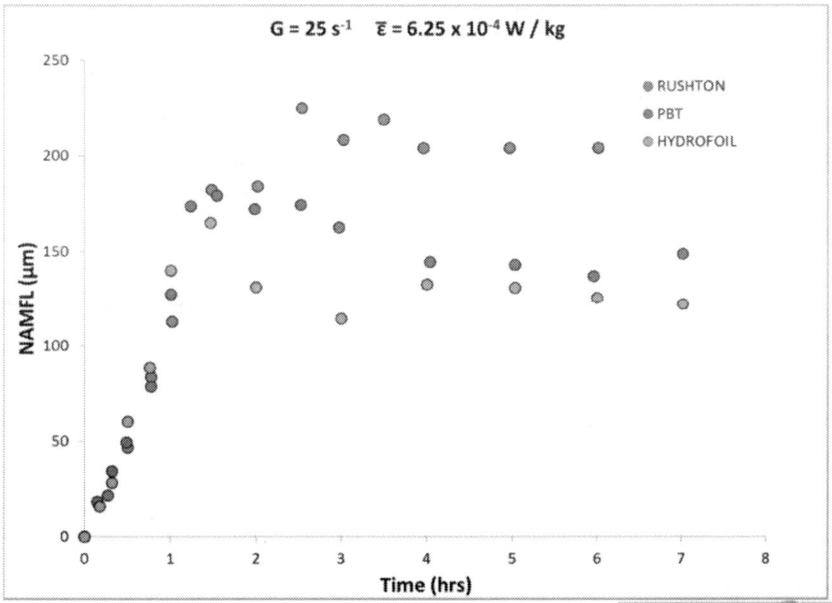

PHILADELPHIA MIXING SOLUTIONS
MIXING SOLUTIONS LIMITED

NAMFL vs. TIME

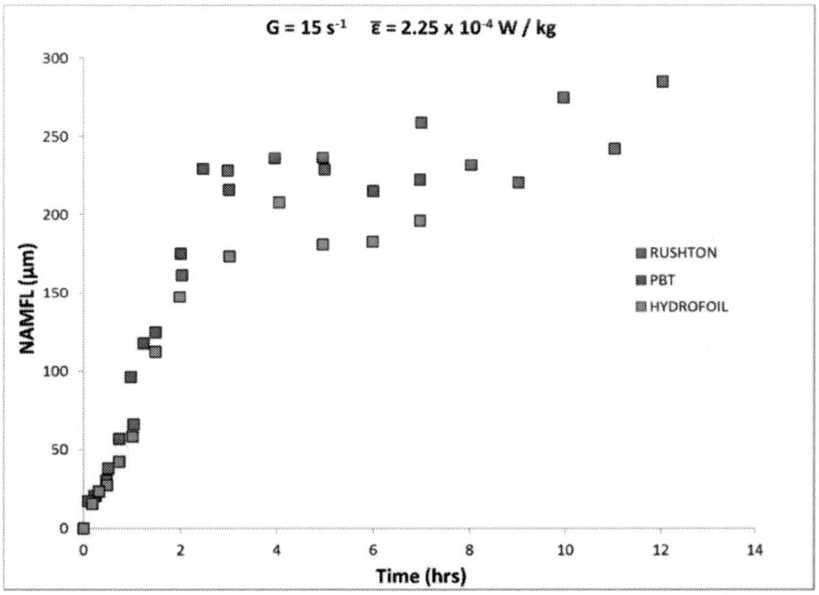

2016 Process Development Symposium | Providence, RI| June 7-9, 2016

PHILADELPHIA MIXING SOLUTIONS
MIXING SOLUTIONS LIMITED

Scaling the LanzaTech Carbon Recycling Process

Gillenwater, T., LanzaTech

Originally presented on: 6/9/2016 10:10:00 - 10:40:00

The need for long term solutions to meet the world's increasing demand for energy and chemicals as well as to reduce greenhouse gas emissions is driving the rapid development of processes to meet this challenge. LanzaTech's innovative gas fermentation technology provides an economic way to recycle the carbon in waste gases using a renewable microbial biocatalyst. Our robust gas fermentation process can use a wide range of feedstocks such as waste gases from steel mills & refineries and syngas produced from biomass or MSW. Carbon that would otherwise be emitted can be captured and recycled with the LanzaTech process to produce chemicals and fuels. A key differentiator of LanzaTech's gas fermentation technology is the ability to use different families of biocatalysts to produce a wide product slate from the same feedstock. LanzaTech's fermentation technology has been extended to methane conversion with an ARPA-E REMOTE project focused on enabling small-scale deployment so that stranded or associated natural gas can be captured to produce alkanes or lipids which can be upgraded to hydrocarbon fuels.

Defeating Scale-up with Process Intensification

Terry Marker, Martin Linck*, Pedro Ortiz-Toral, Jim Wangerow
2016 AIChE Process Development Symposium, Providence, RI, June 7-9

Gas Technology Institute
1700 S Mount Prospect
Des Plaines, IL

Disclaimer

 2

The Challenge – Mobility for the Entire World

gti

- Global emissions of greenhouse gases continue to rise rapidly as economies like those of China, India, and Africa approach a first-world standard of living
- Practical, affordable fuels are a high priority, but if they are derived from fossil sources, the damage to the global climate may be severe and irreversible
- Liquid hydrocarbons comprise the best available transportation fuels: they have an unmatched power density and near-universal market penetration

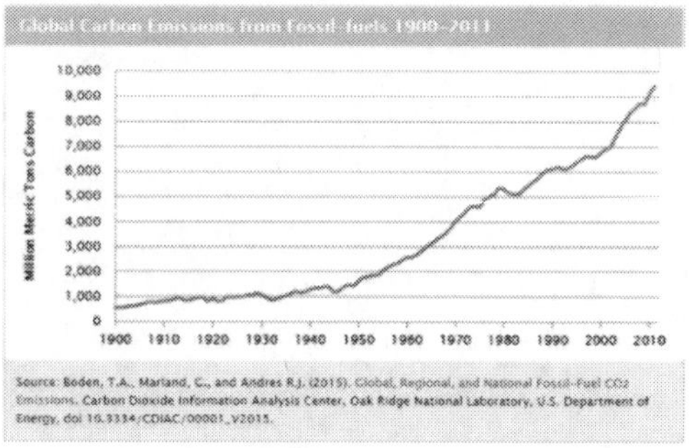

https://www3.epa.gov/climatechange/ghgemissions/global.html

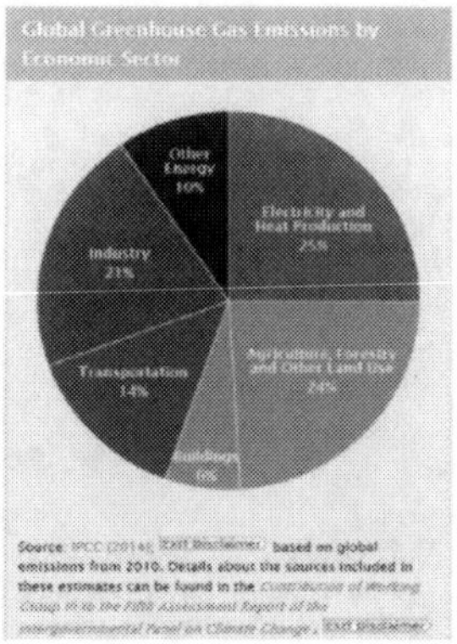

The Approach – Fungible Hydrocarbon Fuels from Biomass

gti

- Other renewable technologies (wind, solar) are able to supply heat and power, but they are a poor fit for the transportation sector.
- Rather than changing the fuels associated with airplanes, trucks, trains, and passenger vehicles, let's derive the best possible fuels from renewable feedstocks: lignocellulosic wood, agricultural wastes, algae, etc.
- There is enough biomass available to supply a significant fraction of the required liquid hydrocarbon transportation fuel the world needs - even the United States could meet 30% of its needs this way[1]
- The developing world has significant biomass resources, but they are available on a distributed basis, and we need to be able to convert them using affordable facilities that scale with the available supply - if it takes a billion dollars to build an economically-viable conversion facility, the facility may never be built
- **Can we apply the principles of process intensification to develop a biomass-conversion approach that produces drop-in liquid hydrocarbon fuels and is highly flexible and deployable?**

1. "Biomass as a Feedstock for a Bioenergy and Bioproducts Industry: The Technical Feasibility of a Billion-Ton Annual Supply," Perlack, et al., 2005

Introduction to IH2®

gti

- IH2® technology is a continuous catalytic thermochemical process that provides a very cost-effective route to produce fungible liquid hydrocarbon transportation fuels from renewable resources.
- Fully profited manufacturing costs are currently estimated at $1.50-$3.50/gallon drop-in transportation fuels, without RIN credits, from $75/ton dry biomass (2012 dollars) at 2000 ton/day dry feed scale (standalone facility) on the basis of FEL-2 capital costs (+/-30%).
- The IH2® process converts virtually any type of non-food biomass feedstock—such as wood, agricultural residues, algae, aquatic plants, and cellulosic fractions of municipal waste—to gasoline, jet and diesel range hydrocarbon fuels and/or blend stocks.
- IH2® technology has been licensed to CRI Catalyst Company (Shell Subsidiary) for exclusive worldwide deployment.

IH2® RENEW. REFINE. REFUEL.

Commercialization Partner

CRI™ DELIVERING INNOVATION

http://www.cricatalyst.com/en/catalysts/renewables.html

How does the IH2® process work?

gti

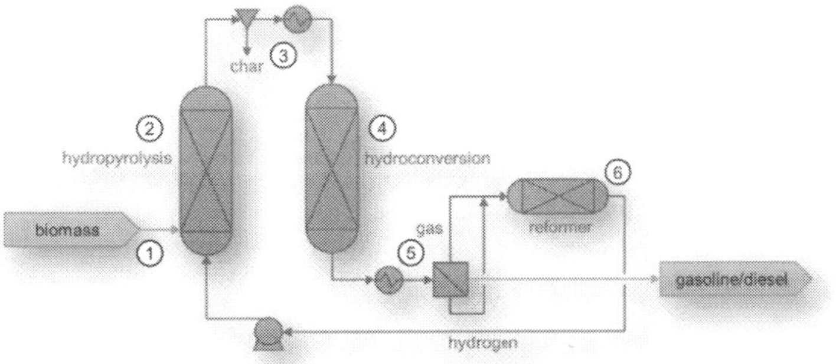

(1) A lockhopper is used to introduce the biomass into a fluidized bed hydropyrolysis reactor.

(2) The hydropyrolysis reactor operates ~400°C and 2-3 MPa of H_2 pressure and contains a bubbling bed of catalyst where biomass is devolatilized, oxygen is converted to H_2O and CO_x, and light hydrocarbon vapor product, C_1-C_3 gases, and char are formed.

(3) Char is continuously removed from the hot vapor in a cyclone followed by a hot filter.

(4) Hydropyrolysis vapors enter a hydroconversion fixed bed reactor with catalyst at ~ 400°C and remaining oxygen and heteroatoms are removed.

(5) Hydrocarbon products and water are then recovered from the product gas and separated into gasoline and diesel products.

(6) Light C_1-C_3 gas is circulated to an integrated steam reformer and pressure swing adsorption unit to produce H_2 that is circulated back to the hydropyrolysis reactor using a compressor.

Why is $IH^{2®}$ technology a breakthrough in biomass to fuels conversion?

Fundamentally, GTI's integrated hydropyrolysis and hydroconversion process (IH^2) represents a breakthrough in biomass conversion to drop-in transportation fuels because it

- directly removes oxygen from biomass using a reducing hydrogen-rich atmosphere and proprietary catalyst
- integrates biomass devolatilization, deoxygenation, and hydrogenation into a single reactor – significant process intensification
- produces a clean, oxygen-free liquid hydrocarbon product (gasoline, diesel, jet fuel)

This represents a major advancement over conventional pyrolysis approaches that produce a thick, black, oxygen-containing pyrolysis oil that is incompatible with today's transportation fuel infrastructure.

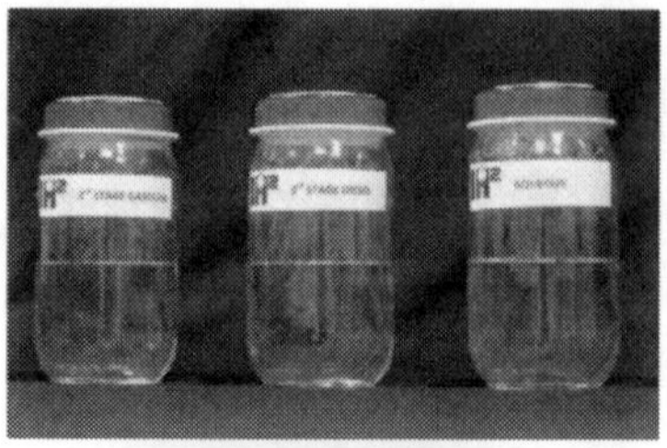

a)

b)

Figure: a) $IH^{2®}$ process products: gasoline (left), diesel/jetfuel range (middle), water (right)
Figure: b) typical oxygen-containing pyrolysis oil/biofuel product.

What are the implications of the $IH^{2®}$ process on biofuels?

- Production of "drop-in" transportation fuels from biomass at a reasonable cost
- Process is relatively simple – small number of steps
- Simplicity of plants allows for commercialization at smaller scales than competing approaches (i.e. Gasification – Fischer Tropsch)
- Potential for small-scale deployment could enable widespread application in distributed markets
- Feedstock flexibility makes process attractive across geographies

> $IH^{2®}$ technology is the first process capable of making drop-in hydrocarbon fuels at prices approaching those of fossil fuels

How Does IH2® Technology Compare?

Table: Comparison of biomass to liquid fuels conversion processes.

	Fast Pyrolysis	Pyrolysis / Upgrading	Hydropyrolysis / Hydroconversion
Product properties	poor	excellent	excellent
Hydrogen required	none	3-4%	none
Capital costs	medium	high	medium
Hot filtering	difficult	difficult	easy
Heat of reaction	endo (300 J/g)	endo / exo	exo / exo
Integrates with upgrading	no	no	yes
Transportation costs	medium	high	low
Char produced	no (burned)	no (burned)	yes

 9

IH2® Process Long-Term Operation

- Approaching 6000 hours of IH2® process pilot plant operation on various softwood and hardwood feedstocks, as well as corn stover
- Long-term tests validated catalyst life, provided process economic model inputs, and produced liquid product for fuel certification tests

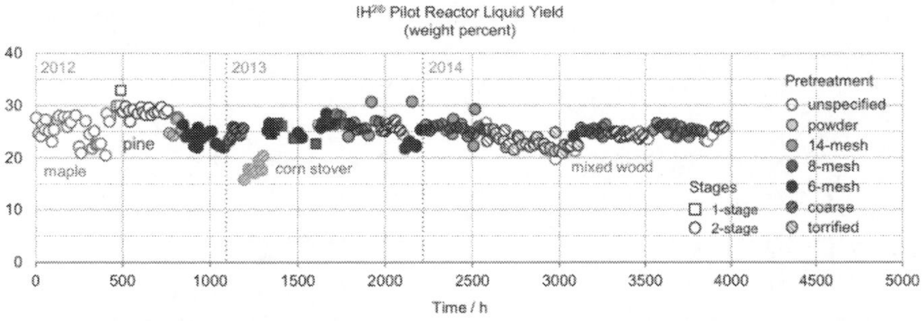

Figure: Liquid yield versus time on stream from IH2 pilot scale reactor during feedstock testing 2012-2014.

IH2® Process Feedstocks

Table: Composition of various feedstocks tested in the IH² process on a moisture and ash-free basis.

Feed	Wood	Lemna	Micro Algae[1]	Macro Algae[2]	Bagasse	Corn Stover
wt% C	49.7	46.3	43.1	34.0	43.1	40.2
wt% H	5.8	5.8	6.1	4.43	5.0	5.0
wt% O	43.9	35.7	20.4	23.6	35.3	35.7
wt% N	0.11	3.7	6.5	4.6	0.34	1.0
wt% S	0.03	0.3	0.7	1.9	0.10	0.05
wt% ash	0.5	8.2	23.1	29.4	16.2	18.1
H/C mole ratio	1.40	1.50	1.70	1.56	1.39	1.49

Table: Product yields of various feedstocks tested in the IH² process on a moisture and ash-free basis.

Product	Wood	Lemna	Micro Algae[1]	Macro Algae[2]	Bagasse	Corn Stover
C4+ liquid / wt%[3]	25-28	30	46	35	30	21
C4+ liquid / (gal/ton)[3]	85	100	157	119	100	67
Oxygen content / wt%	< 0.4	< 0.4	< 0.4	< 0.4	< 0.4	< 0.4
Total acid number	< 0.05	< 0.05	< 0.05	< 0.05	< 0.05	< 0.05

[1] Aquaflow, [2] Blue Marble, [3] moisture and ash-free basis (MAF)

IH2® Process Yields

Table: Typical IH² typical performance results for wood feedstock on a moisture and ash-free basis.

Output	Wood
C_{4+} liquids, wt%	26
Oxygen in liquid, wt%	< 1
Gasoline, wt%	16
Diesel, wt%	10
Char, wt%	13
CO_x wt%	17
C_1-C_3, wt%	13
Water, wt%	36
H_2 uptake, wt%	4.6
H_2 make-up, wt%	0
Ammonia, wt%	0.18
η_{carbon} liquids, %	46
η_{carbon} char, %	25
η_{carbon} process, %	71
$\eta_{thermal}$ liquids, %	57
$\eta_{thermal}$ char, %	24
$\eta_{thermal}$ process, %	81
$\eta_{thermal}$ system, %	60

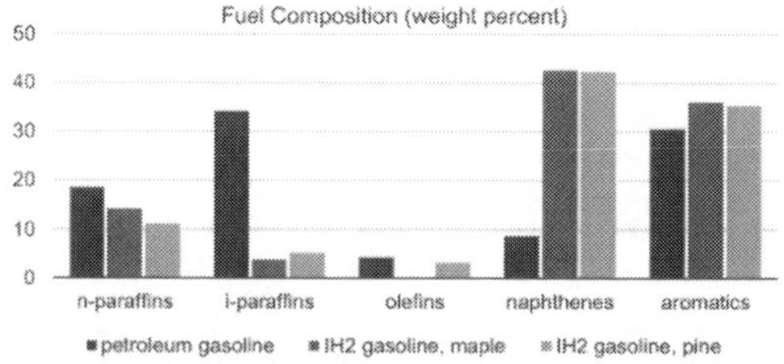

Figure: Comparison of hydrocarbon types in typical petroleum gasoline and from IH² gasoline product using maple and pine feedstocks.

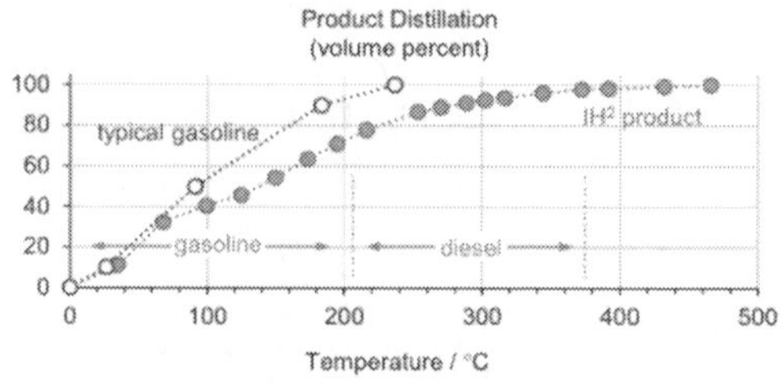

Figure: IH² product distillation volume as a function of temperature compared to a typical gasoline.

$IH^{2®}$ Product Quality

North American Product

Gasoline (from wood)
- R100 meets ASTM D-4814-10b
- Octane ~ 86
- Sulfur < 15 ppm
- Fully renewable premium RBOB
- E10 blend anticipated

Diesel (from wood)
- R100 Meets ASTM D-975-11
- Cetane number 44
- Sulfur < 10 ppm

Drop-in Quality Fuel

European Product

Petrol (from wood)
- EN 228: Petrol – Jan 2009
- RON/MON 87/xx vs 95/85
- Sulfur ~ 10 ppm

Diesel (from wood)
- EN 590:2009+A1:2010
- Cetane number 48
- Lubricity
- Sulfur < 15 ppm

High Quality Blend Stock

$IH^{2®}$ technology transitioning from R&D to commercial deployment

Year	Stage	Feed Capacity	Output Capacity	Description / Status
2009	Proof of Principle	0.5 kg/d	0.2 L/d	• Led by GTI, Des Plaines, IL • Funded internally • Demonstrated wood to hydrocarbon fuel 
2010	Bench Scale	5 kg/d	2 L/d	• Led by GTI, Des Plaines, IL • Funded by DOE • Demonstrated quality fuel production • Demonstrated wood, cornstover, algae, MSW, bagasse, lemna feedstocks 
2012	Pilot Scale	50 kg/d	22 L/d	• Led by GTI/CRI • Funded by DOE/CRI • System engineered and constructed by Zeton • Installed in Des Plaines, IL • Demonstrated continuous operation, high yield • Completed detail techno-economic study 
2016	Pre-Commercial	5,000 kg/d	2,200 L/d	• Led by CRI • First unit planned for Shell Lab, Bangalore, India • Selected Zeton engineering partner • Completed base design engineering package • Awarded (2) "showcase" licenses, FEL-2 level • Plan to commission in 2016 
2018	Commercial	500,000 kg/d	220,000 L/d	• Led by CRI • Selected KBR as exclusive engineering partner • Design - stand alone units for USGC wood • Completed FEL-2 design • Awarded (3) commercial licenses 

IH2® commercialization plowing ahead, even as oil prices plummet

> "SynSel recognizes the potential of the IH2 technology to monetize residual biomass and cost-effectively deliver renewable hydrocarbon fuels and/or blend stocks at prices competitive with fossil fuels currently. SynSel intends to extend its collaboration with CRI and its partners to implement the IH2 technology on a commercial scale in Norway initially."
>
> *Reuters*, Jul 8, 2015

http://www.reuters.com/article/tx-cri-catalyst-company-idUSnBw235320a+100+BSW20150708

> "Shell India Markets Pvt Ltd. will proceed with the installation of a 5 metric ton per day IH2 technology demonstration plant on the site of SIMPL's new Technology Centre in Bangalore, India."
>
> *Biomass Magazine*, Dec 16, 2015

http://biomassmagazine.com/articles/12677/shell-ih2-technology-demonstration-plant-to-be-built

IH2® Process Summary

- Directly makes gasoline, jet, and diesel fuels
- Yield profile is very stable with respect to scale – large scales not needed for optimization
- Competitive standalone facility economics ($1.50-$3.50/gal minimum selling price, without RIN credits)
- Utilizes virtually any biomass feedstock
- Reduces greenhouse gas emissions (emissions equal to 4%-30% of fossil fuel lifecycle[2])
- Runs at moderate hydrogen pressure (100-500 psi)
- Produces own hydrogen; self-sufficient (utilizes C_1-C_3 gas)
- Avoids typical "bad stuff" (polynuclear aromatics, free radicals)
- Oxygen-free and acid-free liquid hydrocarbon product (total acid number < 0.05)
- Clean water byproduct (no carbon contamination)

But why is it better than competing approaches?

2. "A Preliminary Life Cycle Assessment of Biofuels Produced by the IH2 TM Process," Maleche, et al., 2013

Process Comparison – Unique IH$^{2®}$ Advantages

- Comparison 1: Pyrolysis/catalytic pyrolysis followed by upgrading does not appear to be feasible. Many attempts made, no commercially-relevant successes reported
- Comparison 2: Gasification followed by catalytic synthesis involves severe conditions and extensive gas conditioning requirements. Commercially-viable only if the gasification complex is constructed at the ~4000 TPD scale ($1 billion CAPEX requirement or more)
- The simplicity and flexibility of the IH$^{2®}$ approach are of paramount importance:
  - The process is feasible and practical
  - Attractive process economics (even in the absence of subsidies) at 2000 TPD – considerable potential at smaller scales, as well

IH$^{2®}$ successfully demonstrates process intensification and provides unique renewable transportation fuel solutions

RENEW.
REFINE.
REFUEL.

 17

Contacts

Terry Marker
Institute Engineer, Supply and Conversion Division
Gas Technology Institute
(847) 544-3491
terry.marker@gastechnology.org

Martin Linck, PhD
Sr. Engineer, Supply and Conversion Division
Gas Technology Institute
(847) 768-0598
martin.linck@gastechnology.org

Gas Technology Institute
1700 S Mount Prospect
Des Plaines, IL

'Pilotus Maximus'
Best Practices for Pilot & Demo Programs

Alex Patist, Ph.D.
Sr. Dir. BioProcess Technology
Genomatica, Inc.

PDS Process Development Symposium

2016 Process Development Symposium | Providence, RI| June 7-9, 2016

genomatica

Take home message

1. It's not just about the bug
2. Scale-down !
3. Industrial experience, experience, experience

PDS Process Development Symposium

2016 Process Development Symposium | Providence, RI| June 7-9, 2016

genomatica

Begin with the end in mind!

The most important thing is to know where you are going

3

genomatica

Lesson 1: "begin with the end in mind

Apply modeling and techno-economic analysis from the start to guide <u>organism</u>, <u>fermentation</u>, <u>downstream</u>, and <u>pilot efforts</u>

PDS Process Development Symposium

2016 Process Development Symposium | Providence, RI| June 7-9, 2016

genomatica

Current approaches tend toward three camps

- Biology centric
  - Focused on microbe & synthetic biology
  - Process design, economics, and scalability follows later
- Engineering centric
  - Focused on low-cost manufacturing plant
  - "Square peg (biology) is forced into the round hole (the manufacturing plant)"
- Laboratory centric
  - Focused on developing a lab scale microbe & fermentation process
  - Lacks understanding of scale-down for a robust, scalable technology

Disclaimer: Intentionally oversimplified!

Bioengineering: critical to scalable processes

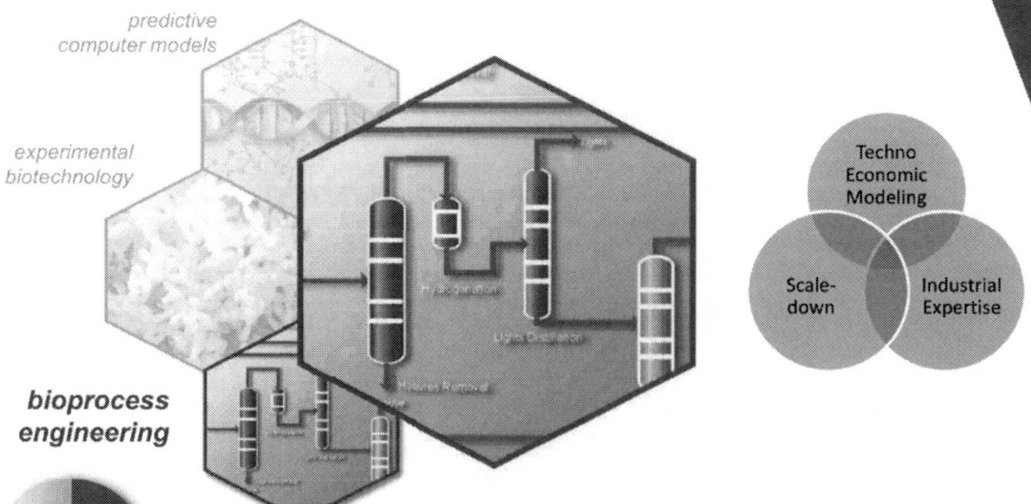

Use techno-economic analysis to drive R&D

core competence that focuses pilot efforts on reducing production costs

Definition: **Combined (simultaneous) process modeling and economic evaluation**

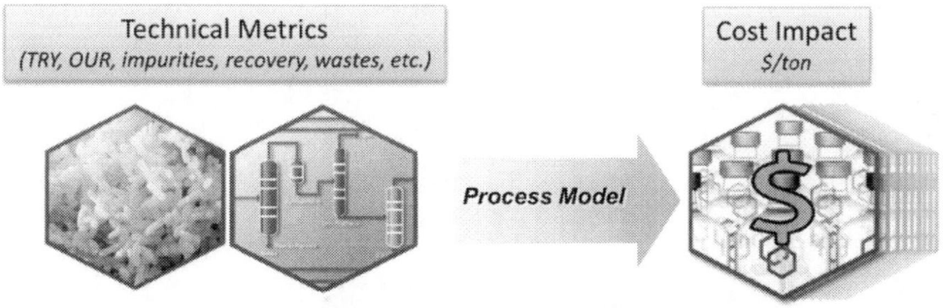

- Identify high-impact areas for optimization
- Assess potential of new opportunities
- Compare economics of design and best alternatives
- Provide metrics for tracking performance

Framework for guiding pilot (R&D) decisions!

genomatica

Production cost nomenclature

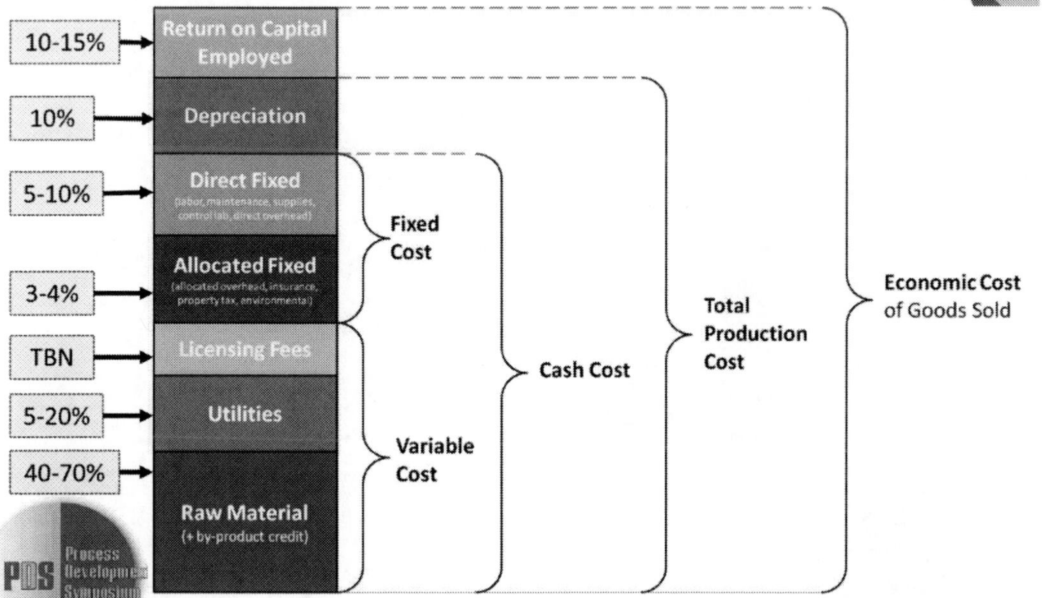

genomatica

2016 Process Development Symposium | Providence, RI| June 7-9, 2016

Identify high impact unit ops (1,4 BDO)

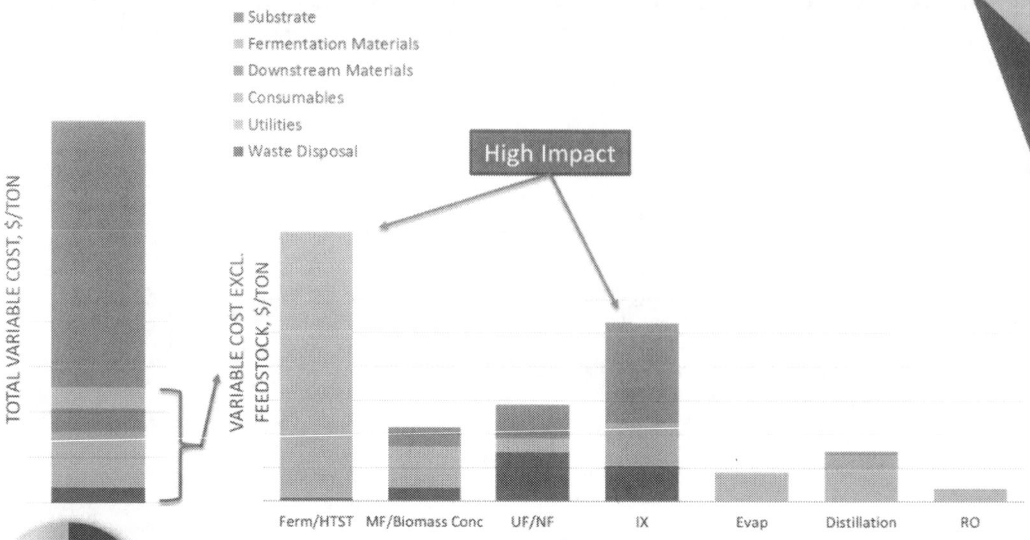

Identify high impact unit ops (cont'd)

Example: Picking a microfiltration design

	Ceramic MF	Polymeric MF
Advantages	• Long membrane life (5 years or more) • Robust; tolerates high temperatures and operating pressures	• Low capital • Low operating costs due to low pumping requirements
Disadvantages	• High capital • High operating costs due to high pumping requirements	• Shorter membrane life (fouling)

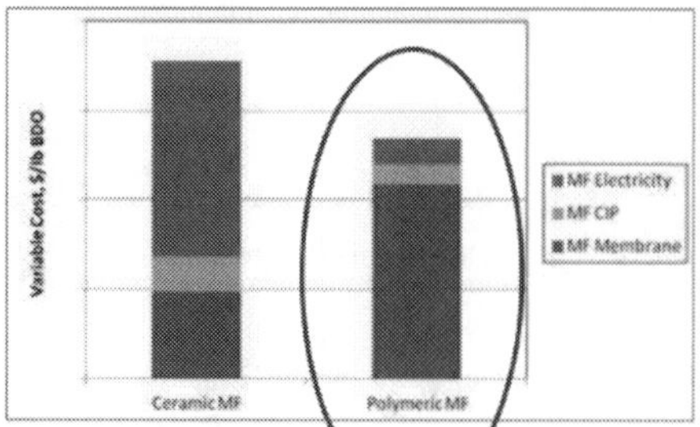

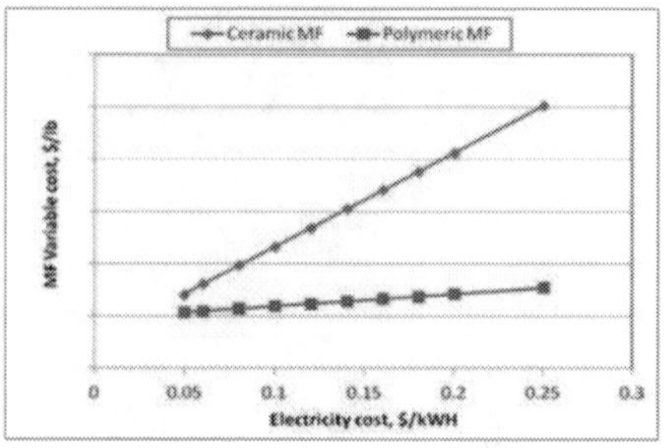

Another high impact area: processing fees !

US corn and raw sugar (NY No. 11) are international benchmarks

- Raw sugar is stored
  - ➢ 96+% sucrose (few impurities)
  - ➢ Unrefined concentrated cane juice is cheaper but carries impurities and only available during the harvesting season
- US corn is stored, traded in US$/bushel (corn bushel equals 56 lbs or 25.5 kg's)
  - ➢ Industrial glucose is a refined syrup 71% solids, ~97.5% fermentable sugars
  - ➢ Glucose cost is (corn price) – (co-product value) + (refining fee)

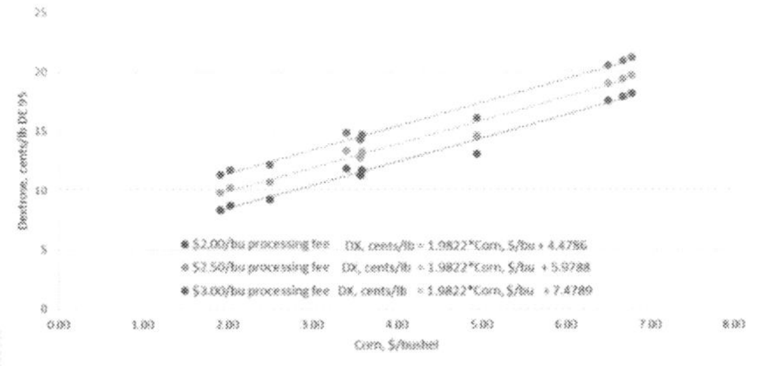

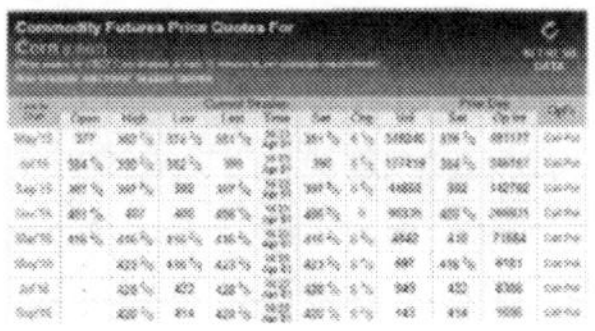

Lesson 2: scale-down !

Mimic big at small !
Do more at small scale and reduce time at the demo

Mimic big at small, <u>not</u> small at big !

"Got a few problems going from lab scale up to full-scale commercial."

Process Development Symposium

genomatica

Scale-down: smaller, cheaper, better, faster

Example: Genomatica's fermentation and process lab

Flexible fermentation capabilities

Iterative experiment design

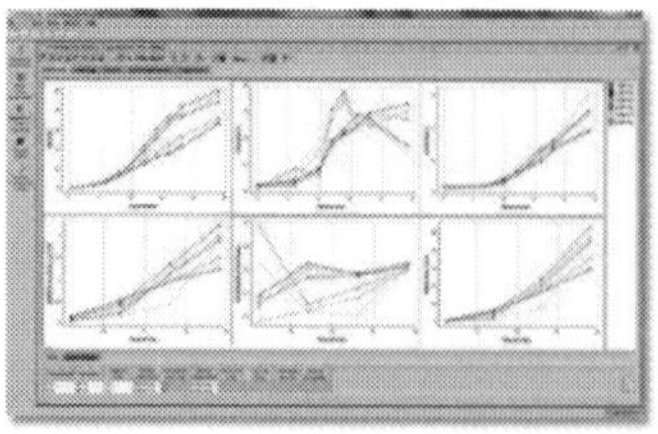

- Quantitative small-scale fermentation design (BioLector®)
- Full capability process lab
  - 50 bioreactors
  - 30L piloting suite to generate kg samples
  - Integrated LIMS
  - Gas mixing, pCO2, wOUR profiles, non-ideal mixing through oscillation control algorithms, integrated bubble column /organism model to understand scale dependent parameters

Process Development Symposium

genomatica

Example: scalable bioprocessing

highly predictable scale-up/down

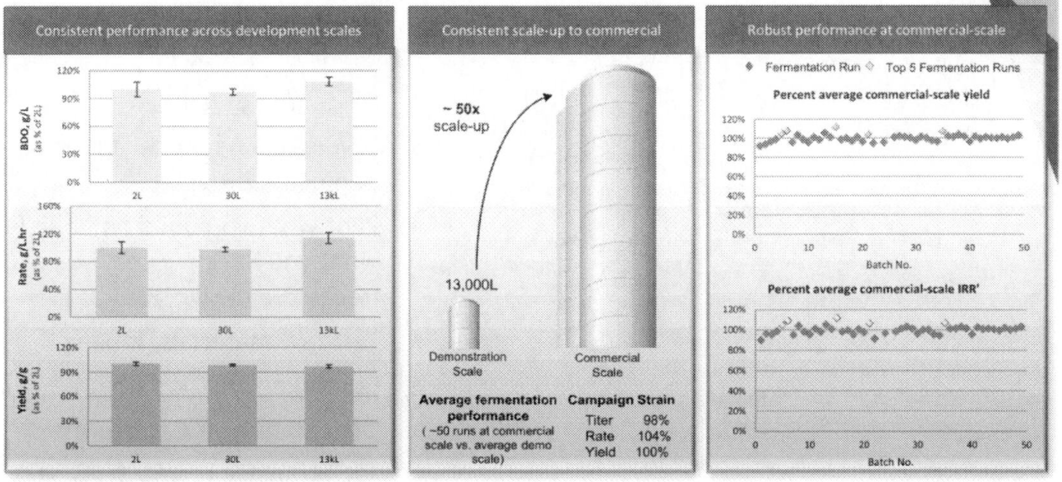

- Fermentation performance across lab, pilot, and demo scales is highly consistent, enabling rapid lab-to-commercial scale development path
- Average commercial-scale performance over ~50 campaign fermentations equivalent to demonstration-scale performance for same strain (+/-2%)
- Low variability in performance across ~50 campaign fermentations, indicates process robustness and predictability

PDS Process Development Symposium

genomatica

2016 Process Development Symposium | Providence, RI| June 7-9, 2016

Lesson 3: industrial experience

Operational issues, such as sterility, fouling and safety are often underestimated at the pilot

PDS Process Development Symposium

genomatica

2016 Process Development Symposium | Providence, RI| June 7-9, 2016

The key to successful deployment: Experience

- Diligent in details
- Leverage a network of flexible demo/toll providers
- Pilot what you build and build what you have piloted, i.e., freeze the design!!
- Test raw materials, fouling
- Contingency/risk, "what if"
- Over staff during start-up
- Tech transfer team

Process Development Symposium

genomatica

How to avoid surprises in technology transfer

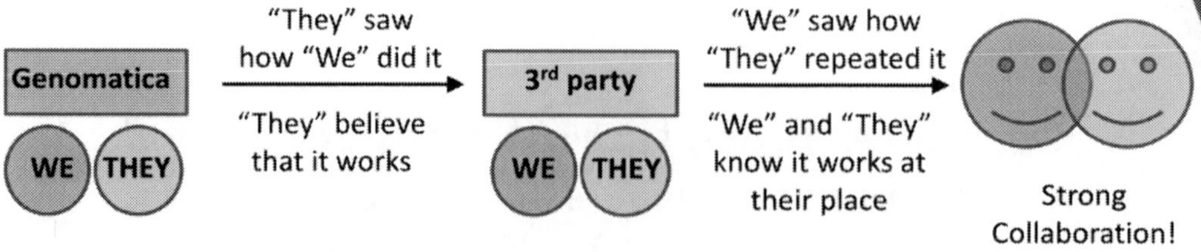

- Shared responsibility (~~hand-offs~~)
- Detailed written protocols (~~verbal~~)
- People move between sites (~~fortresses~~)
- Expect problems – prepare to troubleshoot (~~assume success~~)
- Seamless transfer – over-prepared; boring is good! (~~shortcuts, drama~~)

Process Development Symposium

genomatica

Bioprocess engineering ≠ process engineering

- Carbohydrate chemistry (non-fermentable sugars don't fare well in distillation!)
- Life Cycle Analysis (water, land, biomass waste)
- Impurity profiles (nitrogen!) are managed differently
- Lab scale fermentation process in the plant!
- Consistent feedstock ("Feed the Beast")
- Sterility
- Carbon and nitrogen balances
- Continuous vs. semi-batch
- Weather (external T impact)

PDS Process Development Symposium

genomatica

BDO commercialization journey and success

5 years from concept to commercial production

BDO Producing Organisms – 2008

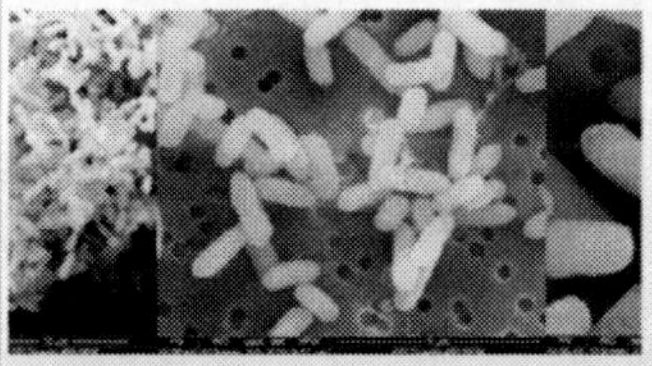

Purified BDO – 2009

3,000 Liter Piloting – 2010

Integrated Demo Plant – 2011

Shipping Tons at a Time – 2011

Commercialization – 2012

PDS Process Development Symposium

genomatica

2016 Process Development Symposium | Providence, RI| June 7-9, 2016

First Bio-1,4BDO plant

2H 2016 start-up, 30kta

Kirkpatrick Award – "the Oscar"

first biotechnology winner of this landmark chemical engineering award

"the most noteworthy chemical engineering technology commercialized in the world in the past two years"

- Award to one company every 2 years
- 80 years of winners have shaped major parts of the chemical industry

Summary

1. Begin with the end in mind:
   Design the process before the 1st gene is cloned!
2. Scale-down experiments:
   Learn more, cheaper, faster and better at smaller scale!
3. Industrial expertise:
   Use experienced individuals who have built/run large scale fermentation plants

PDS Process Development Symposium

genomatica

2016 Process Development Symposium | Providence, RI| June 7-9, 2016

My favorite pilot lessons

- Don't let perfect be the enemy of good enough
- Pilot what you build and build what you have piloted
- In scale-up "Boring is Good"

PDS Process Development Symposium

genomatica

2016 Process Development Symposium | Providence, RI| June 7-9, 2016

genomatica
Thank you
Questions ?
Follow Genomatica:
4757
Alex Patist
Email: apatist@genomatica.com

LCA (Life Cycle Assessment) and Process Development: Start Simple, Think Big!

Richard Helling
The Dow Chemical Company

2016 Process Development Symposium | Providence, RI| June 7-9, 2016

Tridge

Outline

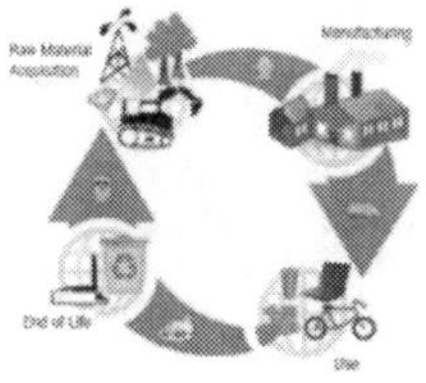

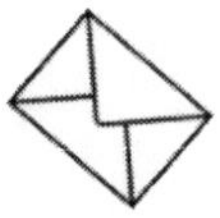

2016 Process Development Symposium | Providence, RI| June 7-9, 2016

$$I = P * A * T$$

2016 Process Development Symposium | Providence, RI| June 7-9, 2016

Planetary Boundaries

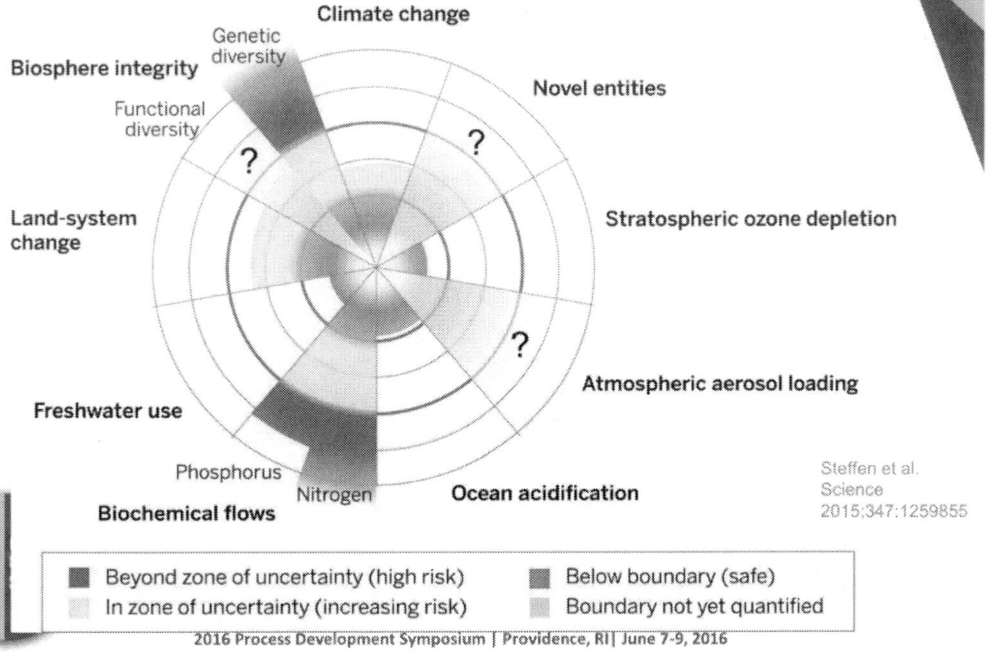

2016 Process Development Symposium | Providence, RI| June 7-9, 2016

LCA improves decisions

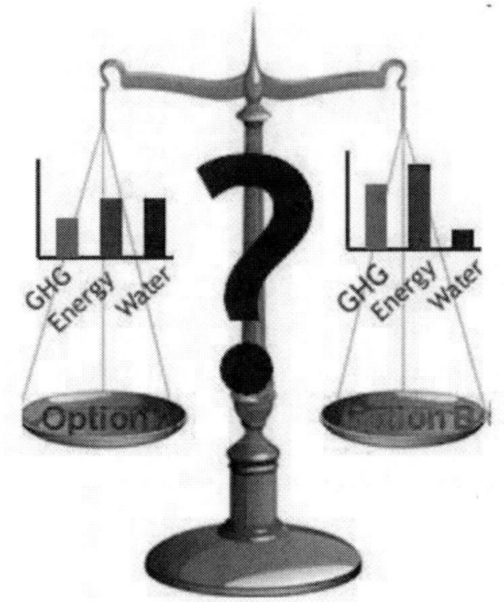

2016 Process Development Symposium | Providence, RI| June 7-9, 2016

LCA framework

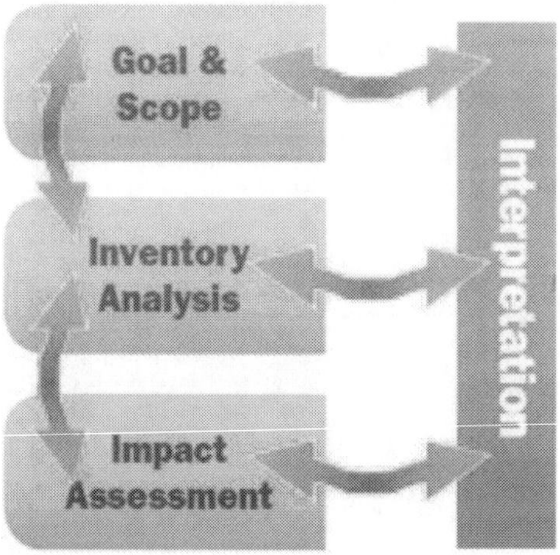

2016 Process Development Symposium | Providence, RI| June 7-9, 2016

Functional unit

Functional Unit: The quantified performance of a product system, so that one can make valid comparisons of them

Poor: A can of paint

Good: The amount of paint required to coat an interior wall surface of 1 ft^2 for 10 years

For PROCESS DEVELOPMENT, per kg production is common

2016 Process Development Symposium | Providence, RI| June 7-9, 2016

Product life cycles

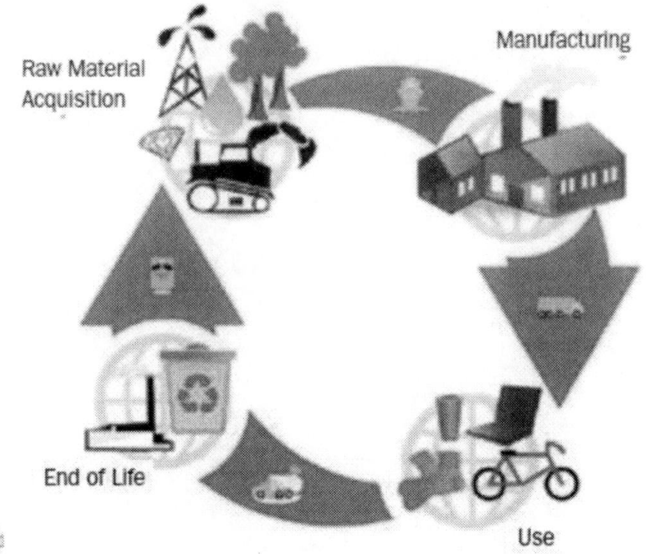

2016 Process Development Symposium | Providence, RI| June 7-9, 2016

Know the inputs & outputs!

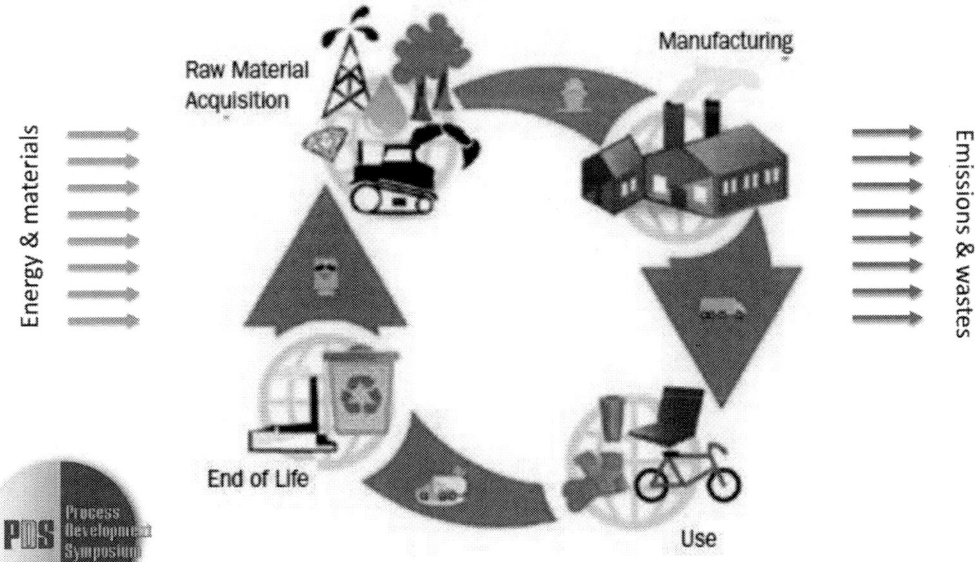

2016 Process Development Symposium | Providence, RI| June 7-9, 2016

2016 Process Development Symposium | Providence, RI| June 7-9, 2016

Calculate potential impacts

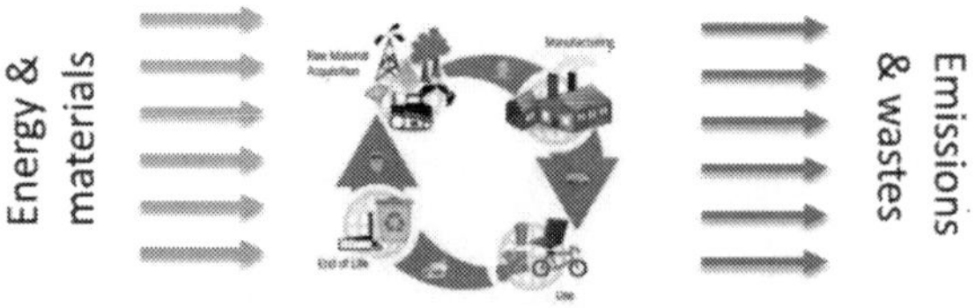

$$\text{Potential Environmental Impact} = \sum_{\substack{\text{all stages}\\ \text{all species}}} \text{Flow} \times \text{Characterization Factor*}$$

*Such as:

- Global Warming Potential (*climate change, carbon footprint*)
- Acidification Potential (*acid rain*)
- Eutrophication Potential (*water pollution, anoxia, dead zone)*
- Photochemical Oxidant Creation Potential (*summer smog*)

2016 Process Development Symposium | Providence, RI| June 7-9, 2016

Tools & Data

2016 Process Development Symposium | Providence, RI| June 7-9, 2016

Tools & Data

Process models, mass basis

Economy-wide
Economic basis

2016 Process Development Symposium | Providence, RI| June 7-9, 2016

Make decisions

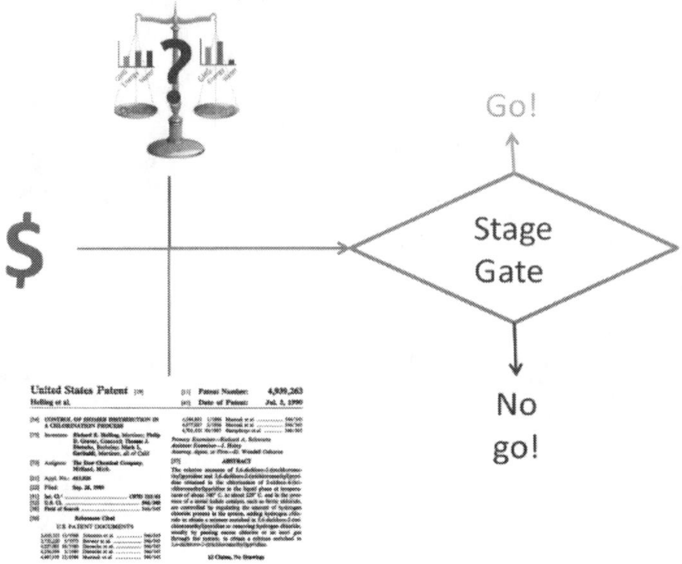

Case study: PO

2010 award: The HPPO process…reduces the…use of energy by 35 percent over traditional technologies

"CHPO" : $C_3H_6 + Cl_2 + 2NaOH = C_3H_6O + 2\ NaCl + H_2O$

"HPPO" : $C_3H_6 + H_2 + O_2 = C_3H_6O + H_2O$

<u>Assume</u>: CHPO use 10 lb steam per lb of PO

What can a life cycle look show?

PO –simple analysis

NOT a life cycle assessment, but an illustration of a simple calculation to support life cycle thinking in process development! Results are not suitable for commercial claims!

Simple calculation **suggests** new route **could** bring both upstream benefits from materials and manufacturing benefits from energy!

2016 Process Development Symposium | Providence, RI| June 7-9, 2016

PO –simple analysis

Quantify energy benefit from “back of envelope”:
35% of 10 lb steam at 85% thermal efficiency from natural gas would save 0.6 kg CO_2 emissions per kg PO

Life cycle concern: Do the new raw materials create more burdens per kg than the benefit?
Estimate using ecoinvent data and ideal stoichiometry

Unit ratios:	CHPO	HPPO	GWP, kg CO2eq/kg of raw
propylene	0.72	0.72	1.44
chlorine	1.22		1.3
NaOH	1.38		1.34
hydrogen		0.03	2.24
oxygen		0.76	0.66
GWP inputs	**4.47**	**1.62**	

Simple calculation **suggests** new route **could** bring both upstream benefits from materials and manufacturing benefits from energy!

2016 Process Development Symposium | Providence, RI| June 7-9, 2016

Case study: THF

Idea:
Replace acid and solvent with high-temperature water (HTW)

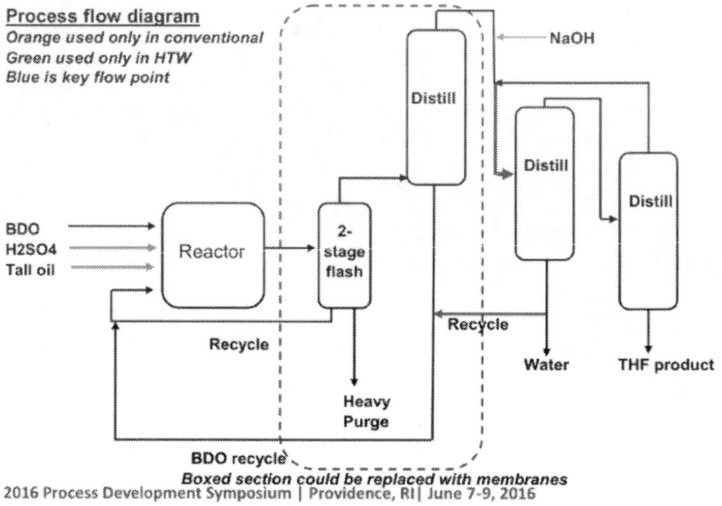

THF: very simple

Quantify material benefit for GWP using published unit ratios and data from ecoinvent: **0.15** kg CO_2eq/kg THF

Life cycle concern: HTW process runs at higher temperature and pressure. Could this burden exceed the benefit from the reduced raw materials?

Unit ratios:	Conventional	HTW	GWP, kg CO2eq/kg of raw
BDO	1.26	1.26	4.21
Sulfuric acid	0.011		0.11
NaOH	0.020		1.34
tall oil	0.027		4.38
GWP inputs	5.43	5.28	

Probably! The GWP benefit is roughly equal to GWP for heating with gas only **0.17** kg of water by 150°C!

THF: What is upstream matters!

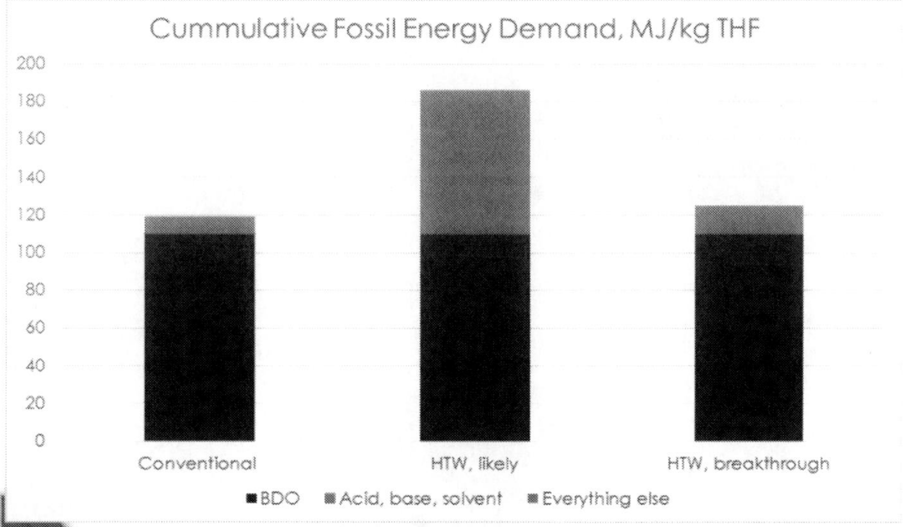

Rigorous ASPENplus modeling confirmed HTW can't beat conventional based on life-cycle energy input

2016 Process Development Symposium | Providence, RI| June 7-9, 2016

Energy as energy and feedstock

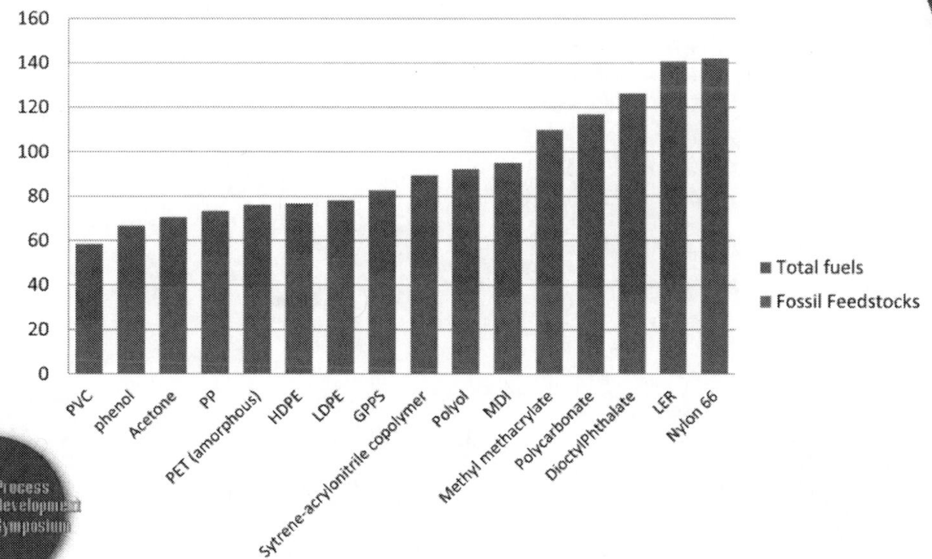

2016 Process Development Symposium | Providence, RI| June 7-9, 2016

Case study: sugarcane PE

vs.

Sugarcane is unlike other feedstocks

2016 Process Development Symposium | Providence, RI| June 7-9, 2016

Sugarcane PE – simple analysis

Estimate maximum benefit for GWP from C content of PE as bio-CO_2 (-3.1 kg CO_2eq/kg) compared to GWP of HDPE from ecoinvent (1.8 kg CO_2eq/kg), or ~**5 kg CO_2eq/kg!**

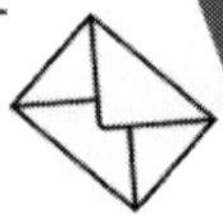

Life cycle concerns: impact of other inputs on GWP, and what of other metrics?

Harder to evaluate the life cycle concerns with the "back of the envelope"!

2016 Process Development Symposium | Providence, RI| June 7-9, 2016

Sugarcane PE: Braskem LCA

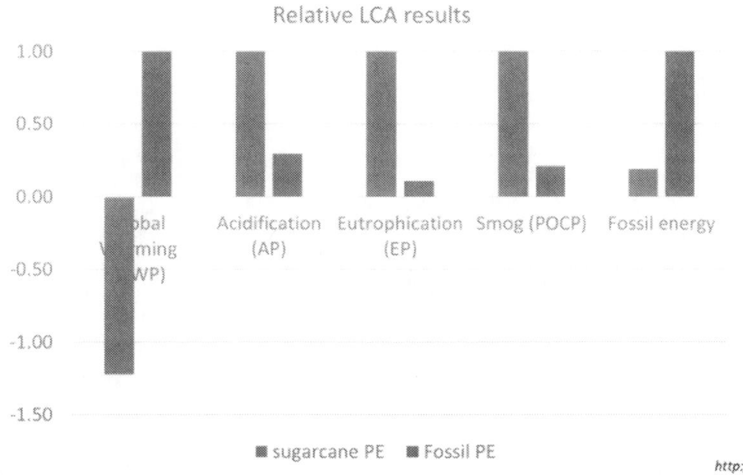

Braskem also analyzed water & land use

http://www.braskem.com/site.aspx/the-life-cycle-assessment-of-its-green-plastic

GWP benefit has trade-offs!

2016 Process Development Symposium | Providence, RI| June 7-9, 2016

Think big!

Product Benefit × A Market Volume = A Potential Market Level Benefit

Hunter, Shawn, and Richard K. Helling "A Call for Technology Developers to Apply Life Cycle and Market Perspectives to Understanding the Potential Environmental Impacts of Chemical Technology Projects", *Industrial & Engineering Chemistry Research*, on-line 26 January (**2015**) DOI: 10.1021/ie504102h

Global benefits of chemical products

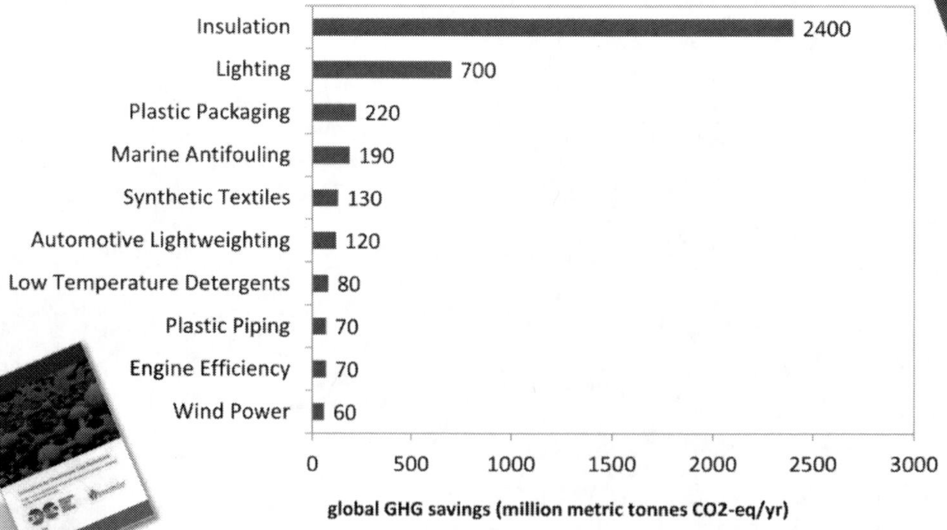

2.1-2.6

Data from: *Innovations for Greenhouse Gas Reductions. A life cycle quantification of carbon abatement solutions enabled by the chemical industry.* ICCA 2009.

GHG savings of chemistry winners

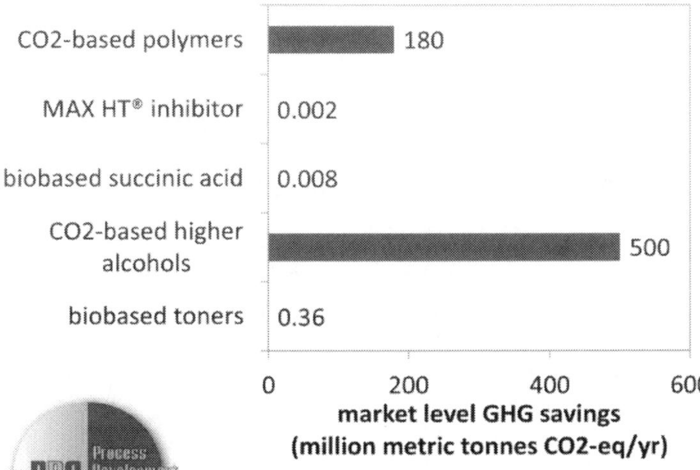

	market penetration assumed	LCA-based GHG estimate?
CO2-based polymers	100%	yes
MAX HT® inhibitor	25%	no
biobased succinic acid	single plant	no
CO2-based higher alcohols	25%	no
biobased toners	25%	yes

Data from: Presidential Green Chemistry Challenge: Award Recipients 1996-2013. United States Environmental Protection Agency, Washington, DC, 2014.

2016 Process Development Symposium | Providence, RI| June 7-9, 2016

Take case studies big

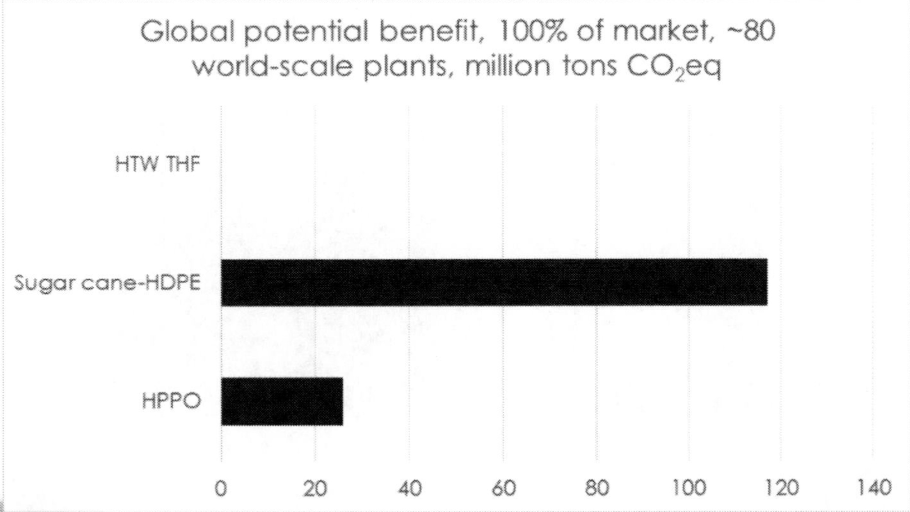

Bio-HDPE looks largest, but would be in fact greatly limited by availability of sugar cane - world HDPE would take ~2X total Brazilian cane!

2016 Process Development Symposium | Providence, RI| June 7-9, 2016

Summary

- Take a life-cycle view – the biggest impacts may not be where you expect!

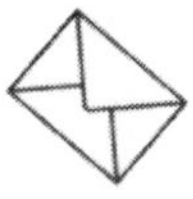

- Start with back of the envelope

- Recognize complexity & trade-offs exist

- Look the biggest opportunities you might influence

2016 Process Development Symposium | Providence, RI| June 7-9, 2016

THANK YOU!

2016 Process Development Symposium | Providence, RI| June 7-9, 2016

An Environmental Life Cycle Assessment Comparison of Single-Use and Conventional Process Technology for Biopharmaceutical Development and Manufacturing

William P. Flanagan, PhD, LCACP
Director, Ecoassessment Center of Excellence
General Electric Company

AIChE Process Development Symposium
Providence, RI

9 June 2016

Imagination at work

Agenda

Biopharmaceutical industry—business drivers

Single-use technologies can enable today's outcomes

Relationship of single-use technology and sustainability

GE's new life cycle assessment study

Insights to guide your single-use journey

Questions

Drug development and manufacturing is a highly complex and competitive industry

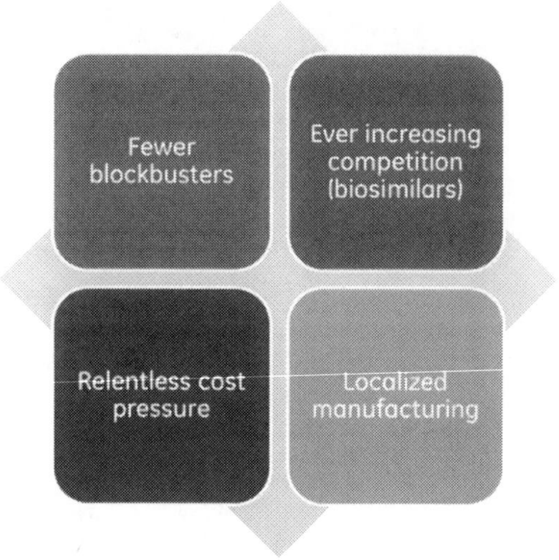

Desired outcomes

Higher productivity, titer/yield

Higher plant utilization

Multi-product, flexible manufacturing

Less variability, failures, and waste

Higher efficiency

Higher manufacturing quality standards

Single-use technologies enabling today's outcomes

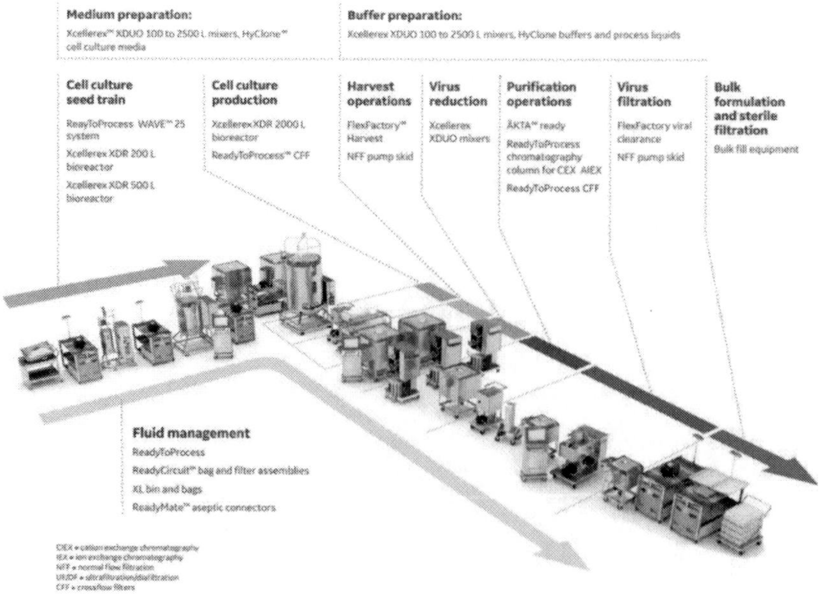

CIEX = cation exchange chromatography
IEX = ion exchange chromatography
NFF = normal flow filtration
UF/DF = ultrafiltration/diafiltration
CFF = crossflow filters

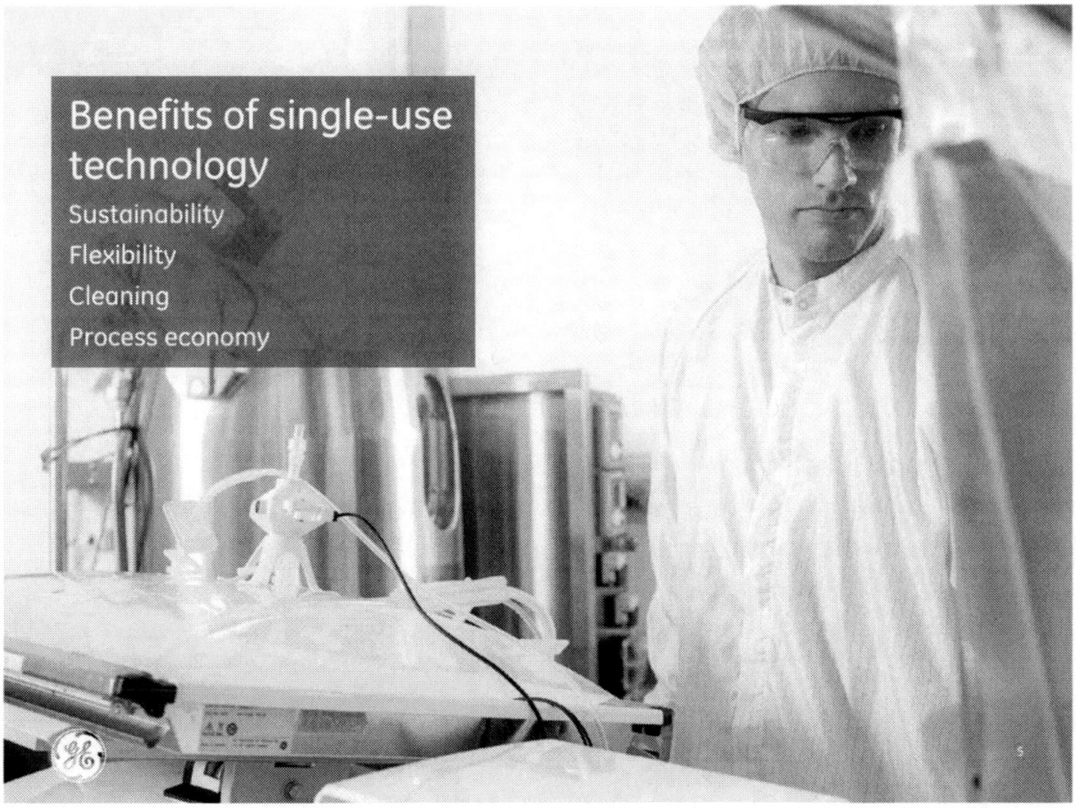

Relationship of single-use and sustainability

Image courtesy of division of Industrial Biotechnology, KTH/Biotechnology

Ecosystem quality | Natural resources | Human health

GE Healthcare FlexFactory™ biomanufacturing operations in Marlborough, MA, USA

- Industry shift towards single-use technology
- Understand environmental trade-offs associated with technology shift
- How can single-use technology help achieve sustainability goals?

Assess overall environmental impact throughout a product or service's life cycle

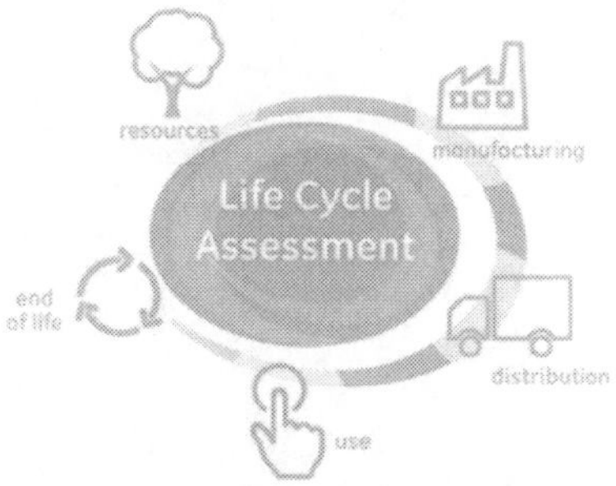

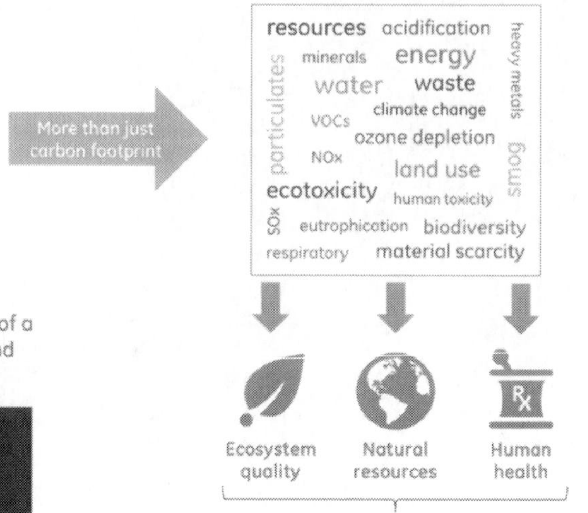

Areas of protection (damage categories)

Understanding the net environmental impact of a product/service across its value chain, how and where to make improvements

- Support decisions
- Evaluate alternatives
- Prioritize opportunities for improvement
- Mitigate environmental issues

2010 to 2012 LCA results

WAVE Bioreactor™ system + ReadyToProcess™ full process train for mAb production

Purpose/driver of LCA

- To compare the potential environmental impacts of multi-use vs. single-use process technology

Results/lessons learned

- Single-use technology exhibits lower environmental impact across the full life cycle
  - reduction of water for injection (WFI), process water, steam
  - less requirement for cleaning and sanitization in place (CIP, SIP)
- End of life impacts negligible compared to use phase and supply chain

Full process train, 2000 L scale

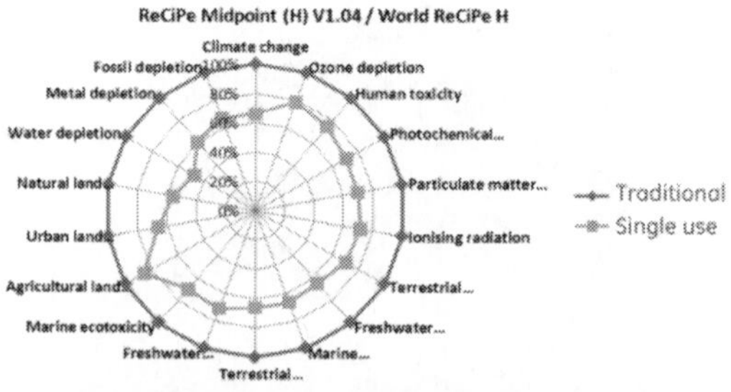

Result was unexpected, counterintuitive, and only accessible through detailed LCA

Pietrzykowksi M. *et al*, An Environmental Life Cycle Assessment Comparison of Single-Use and Conventional Process Technology for the Production of Monoclonal Antibodies, *J. Clean. Prod.* **41**, 150-162 (2013).

Bioprocess evolution requires new insights

- Expansion of single-use capability into new geographies
- Trend towards increased use of facilities with optimized floor plans
- Incorporate broader range of single-use technologies
- Support customer interest and requests for new LCA results

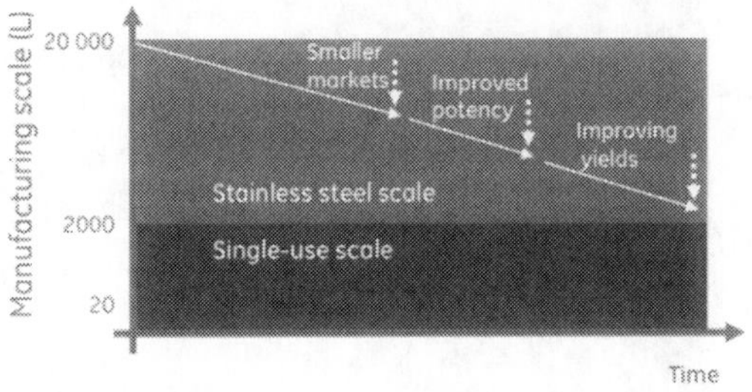

LCA study being updated and expanded

2010–2012 LCA study

mAb process

Stainless steel and single-use retrofit

WAVE Bioreactor™ system and ReadyToProcess™ portfolio

2016 LCA study

mAb and vaccine processes

Stainless steel, single-use retrofit, single-use facilities with optimized floor plan, and hybrid

Xcellerex™ bioreactors, WAVE Bioreactor system, HyClone™ portfolio, ÄKTA™ ready system, ReadyToProcess portfolio

Geography considerations

End-of-life disposal options

LCA study scope, phase I

Image courtesy of division of Industrial Biotechnology, KTH/Biotechnology

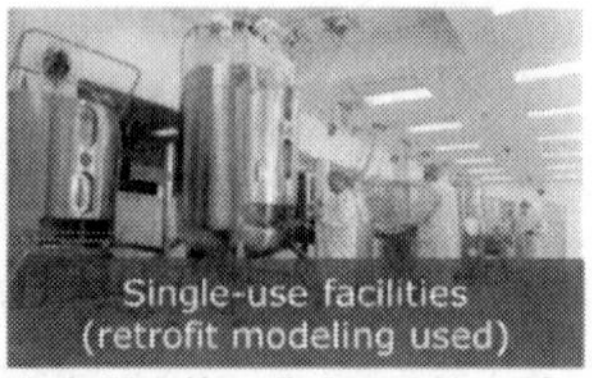

Like the GE Healthcare FlexFactory™ biomanufacturing operations in Marlborough, MA, USA

Like the GE Healthcare KUBio™ facility constructed for JHL Biotech, Wuhan, China

mAb

2 × 2000 L | 2 × 2000 L | 2 × 2000 L

6 g/L titre | 10-batch campaign

Participants: Ecoassessment Center of Excellence at GE, GE Healthcare, Quantis, BioPharm Services, Ltd.

LCA study scope, phase I

Image courtesy of division of Industrial Biotechnology, KTH/Biotechnology

Like the GE Healthcare FlexFactory™ biomanufacturing operations in Marlborough, MA, USA

Like the GE Healthcare KUBio™ facility constructed for JHL Biotech, Wuhan, China

mAb

2 × 2000 L | 2 × 2000 L | 2 × 2000 L

Third party critical review to be performed

Participants: Ecoassessment Center of Excellence at GE, GE Healthcare, Quantis, BioPharm Services, Ltd.

Study scope and system boundaries: traditional bioprocessing

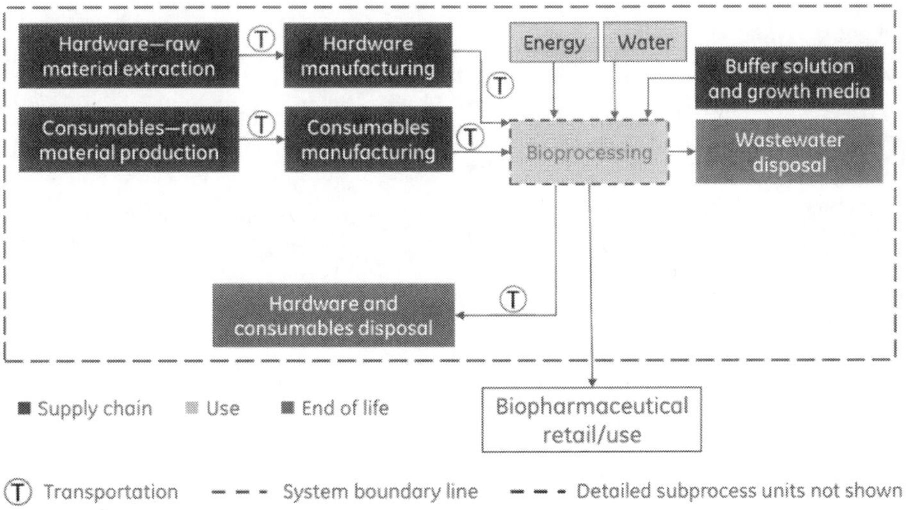

Study scope and system boundaries: traditional bioprocessing

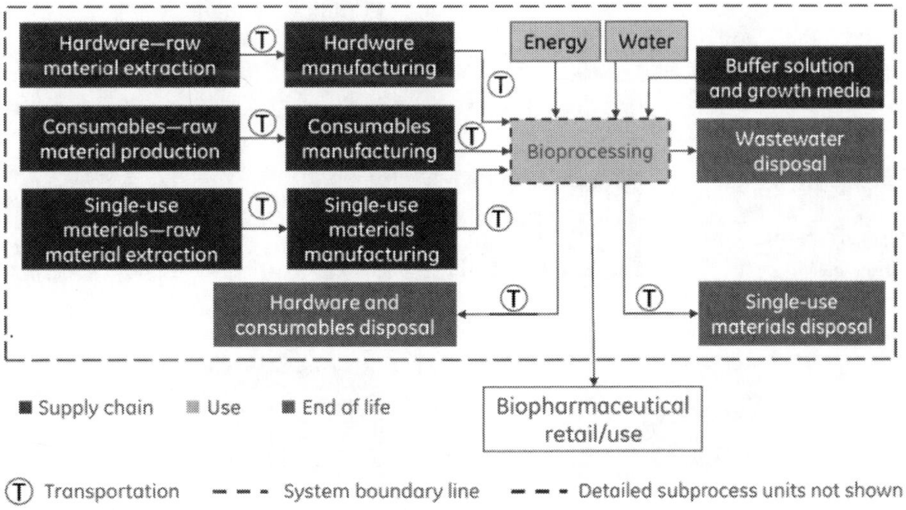

■ Supply chain ■ Use ■ End of life

mAb bioprocessing configurations studied

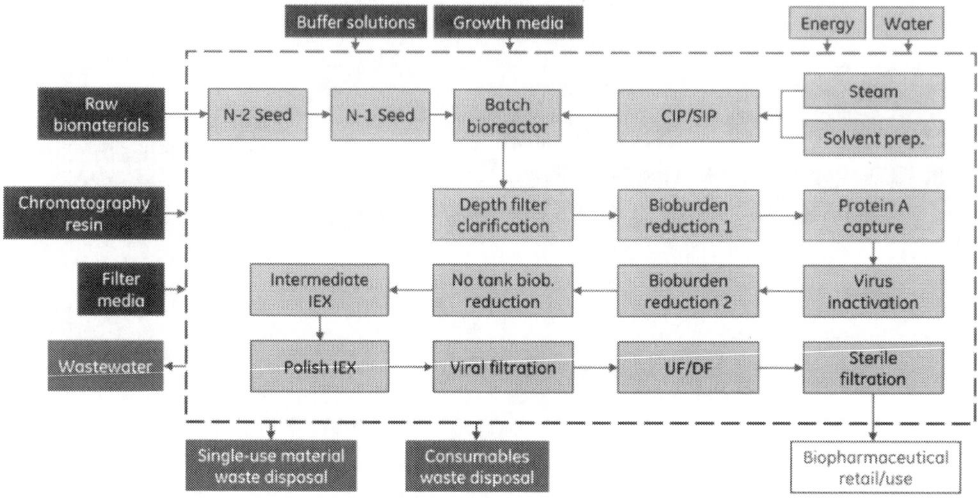

biob. = bioburden, CIP/SIP = cleaning/sterilization in place, IEX = ion exchange chromatography, prep. = preparation, UF/DF = ultrafiltration/diafiltration
Note: General unit operations shown; configuration can change due to scale, product choice, technologies used, etc.

Bioprocess modeling and data collection

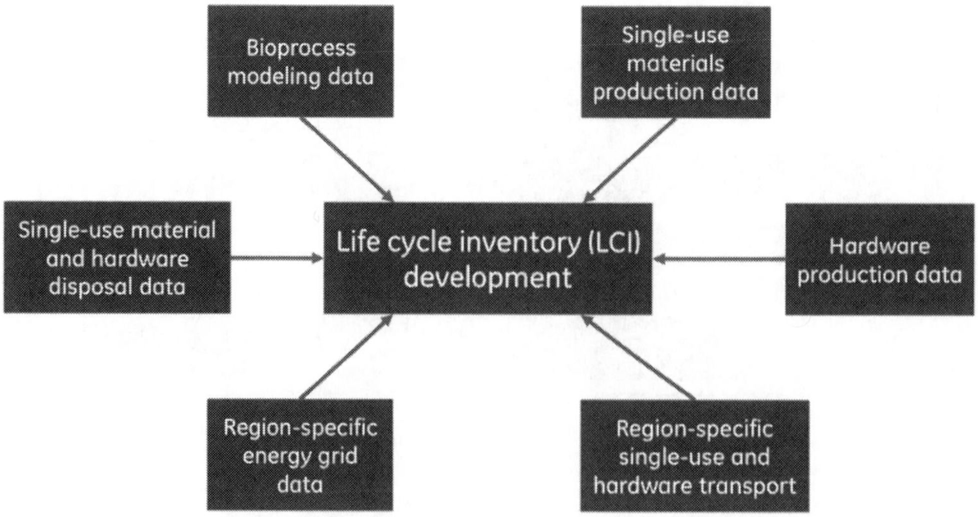

IMPACT 2002+ impact assessment

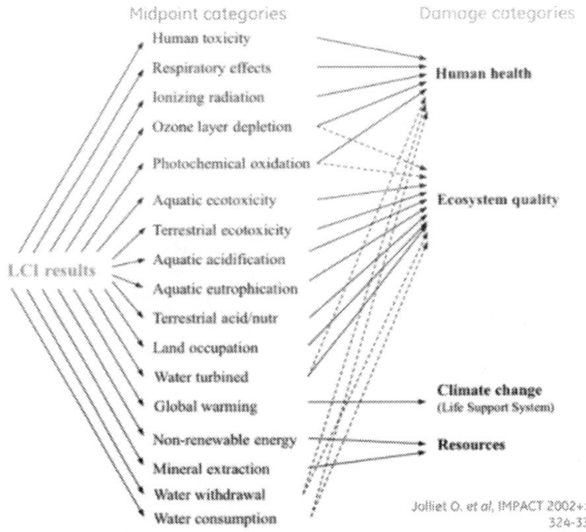

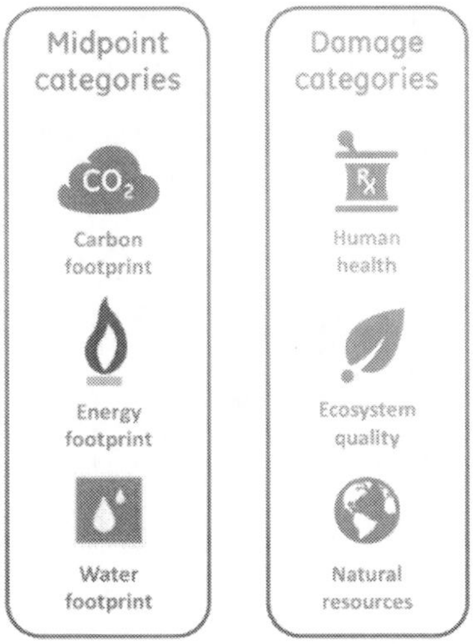

Jolliet O. *et al*, IMPACT 2002+: a new life cycle impact assessment methodology, *Int. J. Life Cycle Assess* 8(6), 324-330 (2003) as adapted by Quantis in version Q2.22 of IMPACT 2002+: User Guide

LCA study results for 2 × 2000 L mAb process

Looking at the LCA results from three different bioprocess perspectives

US-based, medium-sized biotechnology company expanding capacity

Large multi-national biopharmaceutical company expanding global capacity

Large multi-national biopharmaceutical company upgrading existing stainless steel process

Does the increased use of plastics mean my environmental impact will increase?

US-based, medium-sized biotechnology company expanding capacity

30% to 60% reduction in life cycle impact across the five environmental impact categories

Single-use technologies provide a 30% to 60% reduction in total impact across the five environmental impact categories compared with traditional technologies

North American East Coast, 2 × 2000 L processing scale

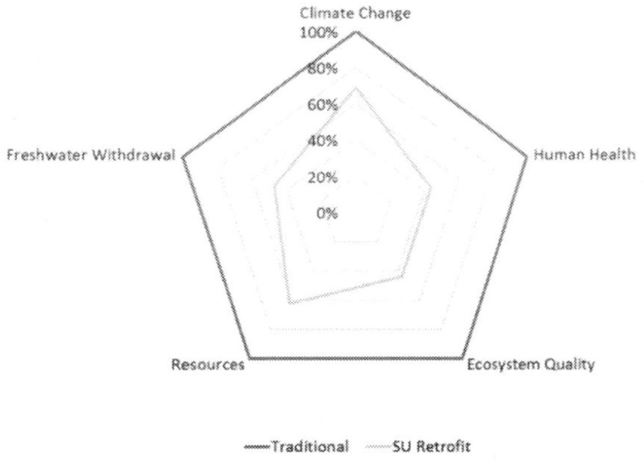

Preliminary results

SU Retrofit = single-use facility (retrofit modeling), Traditional = traditional stainless steel facility

Further reductions in life cycle impact for a single-use facility with optimized floor plan

The process chain for a single-use facility with optimized floor plan shows further savings in all impact categories studied.

North American East Coast, 2 × 2000 L processing scale

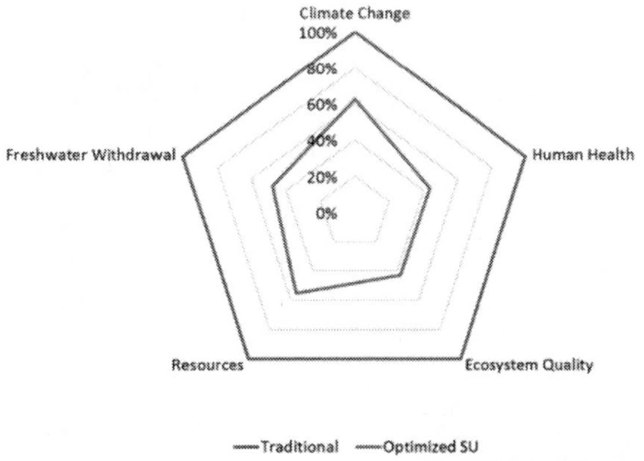

Preliminary results

Optimized SU = single-use facility with optimized floor plan, Traditional = traditional stainless steel facility

Differences in impact during a product's lifecycle

The single-use process scenario has a larger share of impact in the supply chain, while greatly reducing the impact in use

North American East Coast,
2 × 2000 L processing scale

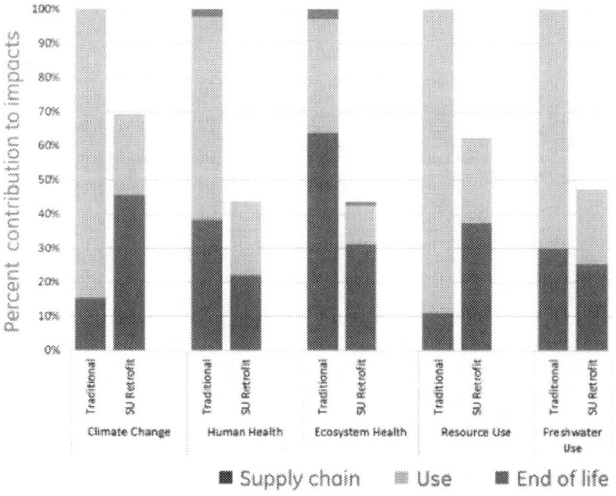

Preliminary results

SU Retrofit = single-use facility (retrofit modeling), Traditional = traditional stainless steel facility

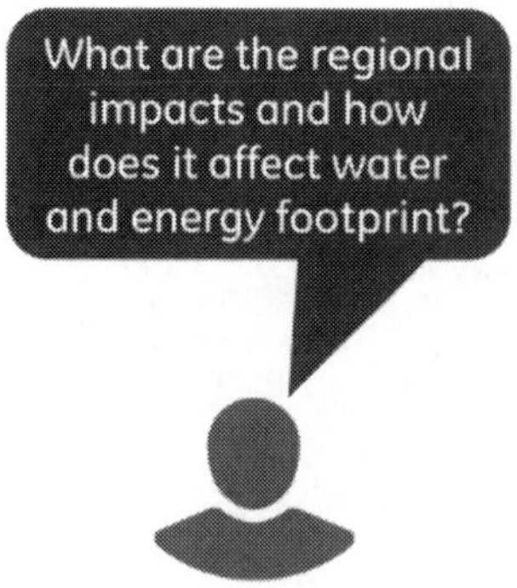

Large multi-national biopharmaceutical company expanding global capacity

The carbon footprint savings vary among the three locations examined

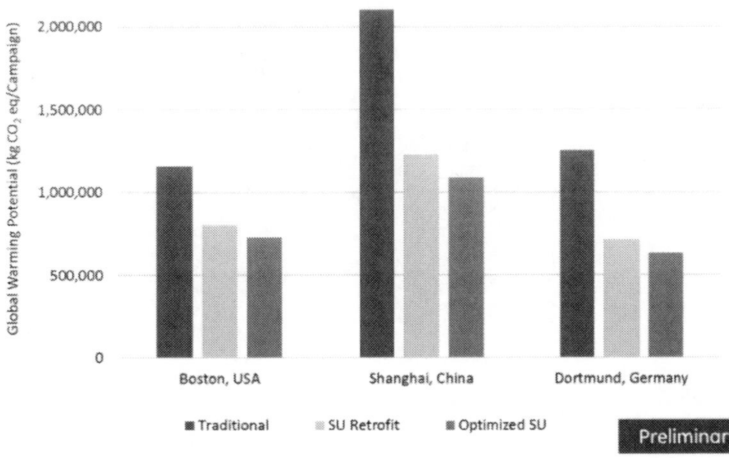

Preliminary results

SU Retrofit = single-use facility (retrofit modeling), Optimized SU = single-use facility with optimized floor plan, Traditional = traditional stainless steel facility

The water use savings vary among the three locations examined

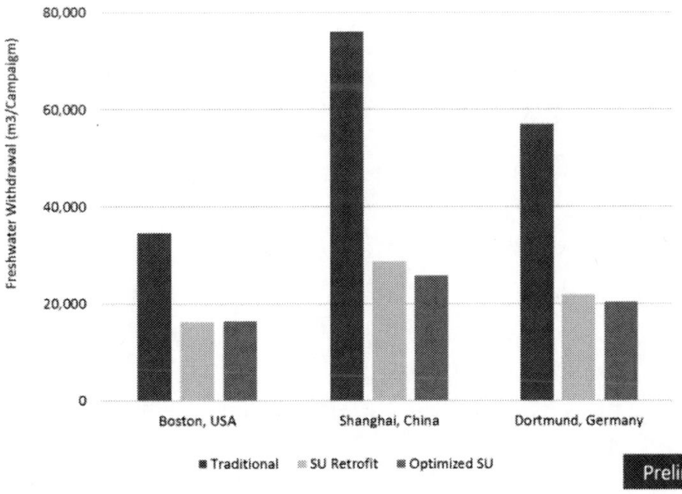

Preliminary results

SU Retrofit = single-use facility (retrofit modeling), Optimized SU = single-use facility with optimized floor plan, Traditional = traditional stainless steel facility

The energy use savings vary among the three locations examined

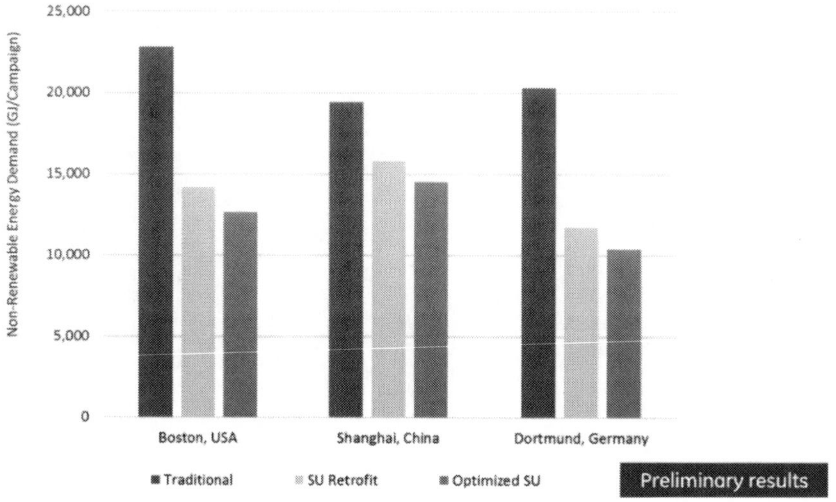

SU Retrofit = single-use facility (retrofit modeling), Optimized SU = single-use facility with optimized floor plan, Traditional = traditional stainless steel facility

Large multi-national biopharmaceutical company upgrading existing stainless steel process

Achieving carbon footprint reduction goals

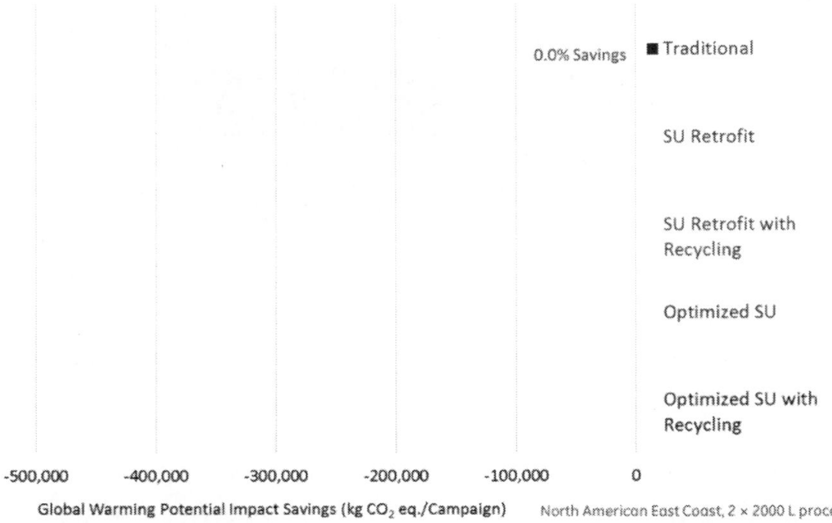

North American East Coast, 2 × 2000 L processing scale

SU Retrofit = single-use facility (retrofit modeling)
Optimized SU = single-use facility with optimized floor plan
Traditional = traditional stainless steel facility

Preliminary results

Achieving carbon footprint reduction goals

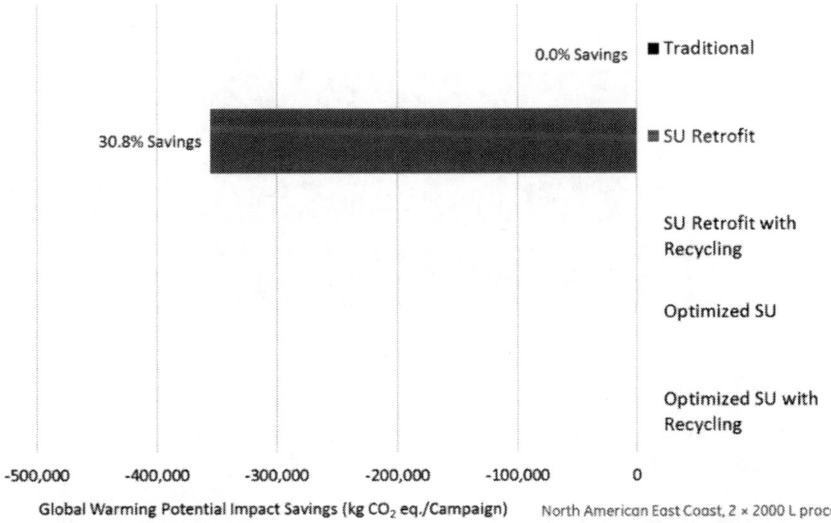

North American East Coast, 2 × 2000 L processing scale

SU Retrofit = single-use facility (retrofit modeling)
Optimized SU = single-use facility with optimized floor plan
Traditional = traditional stainless steel facility

Preliminary results

Achieving carbon footprint reduction goals

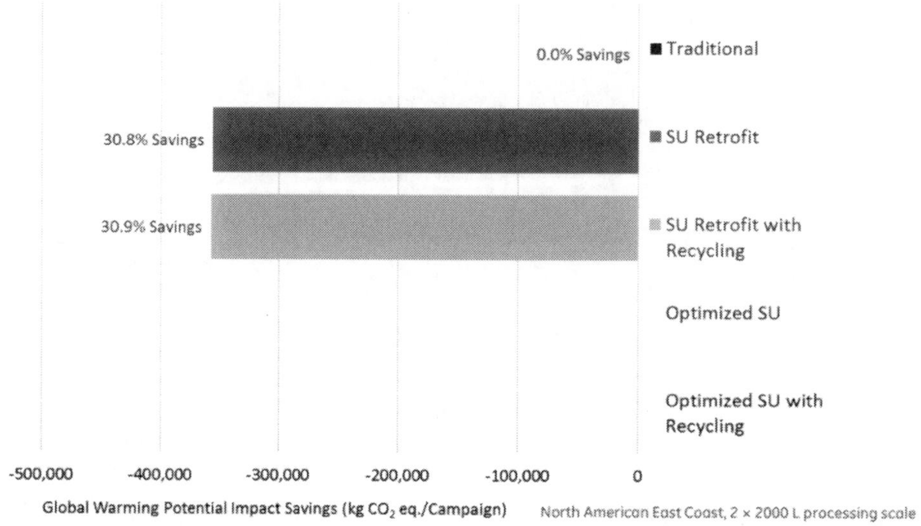

North American East Coast, 2 × 2000 L processing scale
SU Retrofit = single-use facility (retrofit modeling)
Optimized SU = single-use facility with optimized floor plan
Traditional = traditional stainless steel facility

Preliminary results

AIChE Process Development Symposium | Providence, RI | June 7-9 2016 31

Achieving carbon footprint reduction goals

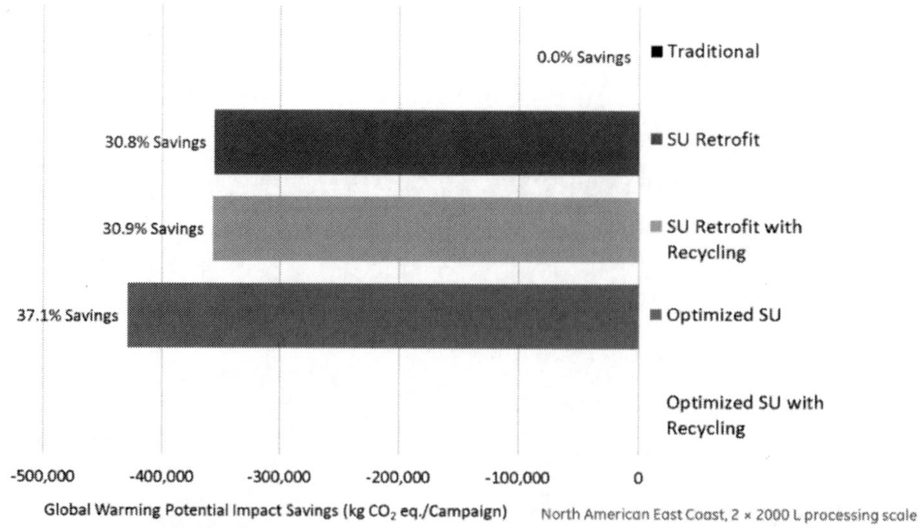

North American East Coast, 2 × 2000 L processing scale
SU Retrofit = single-use facility (retrofit modeling)
Optimized SU = single-use facility with optimized floor plan
Traditional = traditional stainless steel facility

Preliminary results

AIChE Process Development Symposium | Providence, RI | June 7-9 2016 32

Achieving carbon footprint reduction goals

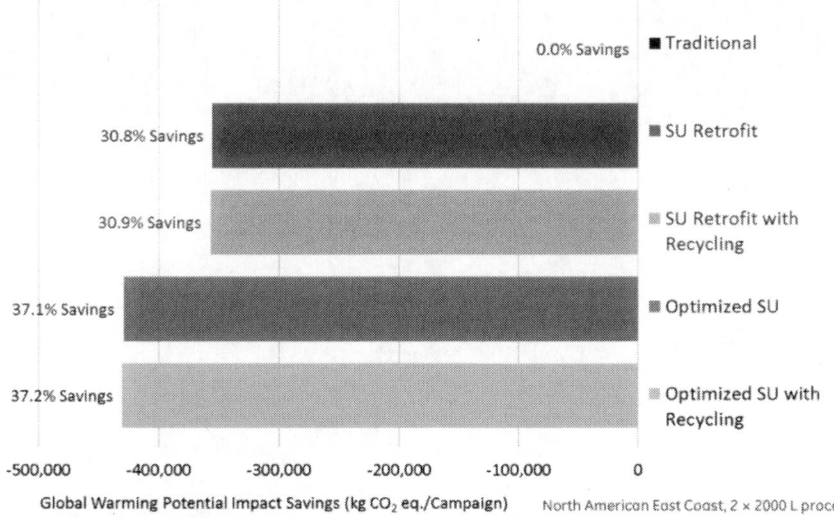

North American East Coast, 2 × 2000 L processing scale

SU Retrofit = single-use facility (retrofit modeling)
Optimized SU = single-use facility with optimized floor plan
Traditional = traditional stainless steel facility

Preliminary results

Summary of LCA study phase I

Single-use technology exhibits lower environmental impact compared with traditional technology

End of life contribution is negligible relative to overall environmental impact

Geographic analysis demonstrates single-use technology exhibits varying levels of environmental impact benefits

Single-use technology can help achieve corporate sustainability development goals

Sustainability profiles are process dependent—
engage our team for the next phase

LCA study scope, phase II

Image courtesy of division of Industrial Biotechnology, KTH/Biotechnology

Like the GE Healthcare FlexFactory™ biomanufacturing operations in Marlborough, MA, USA

Like the GE Healthcare KUBio™ facility constructed for JHL Biotech, Wuhan, China

	Traditional stainless steel facilities		Single-use facilities (retrofit modeling used)	Single-use facilities with optimized floor plan
mAb	200 L		200 L	
	500 L		500 L	
	2000 L	Stainless steel/ single-use hybrid	2000 L	2 × 1000 L
	2 × 2000 L		2 × 2000 L	2 × 2000 L
	4 × 2000 L		4 × 2000 L	4 × 2000 L

Participants: Ecoassessment Center of Excellence at GE, GE Healthcare, Quantis, BioPharm Services, Ltd.

LCA study scope, phase III

Image courtesy of division of Industrial Biotechnology, KTH/Biotechnology

Like the GE Healthcare FlexFactory™ biomanufacturing operations in Marlborough, MA, USA

Like the GE Healthcare KUBio™ facility constructed for JHL Biotech, Wuhan, China

	Traditional stainless steel facilities		Single-use facilities (retrofit modeling used)	Single-use facilities with optimized floor plan
mAb	200 L		200 L	
	500 L		500 L	
	2000 L	Stainless steel/ single-use hybrid	2000 L	2 × 1000 L
	2 × 2000 L		2 × 2000 L	2 × 2000 L
	4 × 2000 L		4 × 2000 L	4 × 2000 L
Vaccine	10 L		10 L	
	200 L		200 L	

LCA model can explore environmental improvement opportunities in stainless steel, single use, or hybrid

What is your scenario?

This is an ongoing, collaborative study addressing customer sustainability questions

What are your key questions relative to single-use and sustainability?

Acknowledgements

BioPharm Services

Adriana Lopes

Andrew Sinclair

Quantis International

Carter Reeb

Jon Dettling

GE Healthcare

Amit Dua

Bill Whitford

Johan Rosenquist

Mia Bennemo

Karolina Busson

Thank you!

Bill Flanagan
Director, Ecoassessment Center of Excellence
General Electric Company

Chair, Board of Directors
American Center for Life Cycle Assessment

UNIT OPERATION-SPECIFIC BREAKDOWN OF DURABLE AND SINGLE-USE TECHNOLOGY ASSUMPTIONS

Unit op. number	Unit op. name	Technologies used for comparison at 2 × 2000 L scale		
		Traditional	Single-use retrofit	Single-use facility with optimized floor plan
1	N-3 seed	SS bioreactor	ReadyToProcess™ WAVE 25 bioreactor system	ReadyToProcess WAVE 25 bioreactor system
2	N-2 seed	SS bioreactor	Xcellerex™ XDR-200 bioreactor	Xcellerex XDR-200 bioreactor
3	N-1 seed	SS bioreactor	Xcellerex XDR-500 bioreactor	Xcellerex XDR-500 bioreactor
4	Production	SS bioreactor	Xcellerex XDR-2000 bioreactor	Xcellerex XDR-2000 bioreactor
5	Clarification	SS filter housing + depth filtration	SU harvest filter system and holder, SU depth filter, and SU polishing filter	SU harvest filter system and holder, SU depth filter, and SU polishing filter
6	Bioburden reduction I	SS filter housing + bioburden filter		
7	Protein A	SS column + protein A resin	ÄKTA™ ready system + AxiChom 450 column, MabSelect SuRe™ resin	ÄKTA ready system + AxiChom 450 column, MabSelect SuRe resin
8	Virus inactivation	SS vessel	Xcellerex XDUO 100 and 1000 L mixers	Xcellerex XDUO 100 and 1000 L mixers
9	Bioburden reduction II	SS filter housing + bioburden filter		
10	Sterile filtration I	SS filter housing + sterile filter	(Peristaltic pump) filter capsule/sterile filtration	(Peristaltic pump) filter capsule/sterile filtration
11	IEX bind-elute	SS column + IEX resin	ÄKTA ready system + ReadyToProcess column prepacked with Capto™ S ImpAct resin	ÄKTA ready system + ReadyToProcess column prepacked with Capto™ S ImpAct resin
12	UF/DF	TFF SS system + UF filter	TFF SS system (UniFlux™) + UF filter (cassettes)	TFF SS system (UniFlux™) + UF filter (cassettes)
13	AIEX flowthrough	SS column + AIEX resin	ÄKTA ready system + ReadyToProcess column prepacked with Capto™ Q resin	ÄKTA ready system + ReadyToProcess column prepacked with Capto™ Q resin
14	Viral filtration	SS filter housing + viral filter	Millipore Pro Magnus Viral filter skid and holder + Viresolve™ Pro	Millipore Pro Magnus Viral filter skid and holder + Viresolve Pro
15	UF/DF	TFF SS system + UF filter	TFF SS system (UniFlux) + UF filter (Millipore Pellicon™ cassettes)	TFF SS system (UniFlux) + UF filter (Millipore Pellicon cassettes)
16	Sterile filtration II	SS filter housing + sterile filter	Peristaltic pump + filter capsule/sterile filtration	Peristaltic pump + filter capsule/sterile filtration

IEX = ion exchange chromatography, op. = operation, SS = stainless steel, SU = single use,
TFF = tangential flow filtration, UF/DF = ultrafiltration/diafiltration
SU retrofit = single-use facility (retrofit modeling), Traditional = traditional stainless steel facility

Including Ecosystem Services for Developing Innovative Processes and Life Cycles

Bhavik R. Bakshi

William G. Lowrie Department of Chemical and Biomolecular Engineering
The Ohio State University, Columbus OH 43210, USA

The Ohio State University

Process Development Symposium, American Institute of Chemical Engineers
Providence, Rhode Island, June 7-9, 2016

Requirements for Sustainable Manufacturing

Efforts toward sustainability should

1. Prevent shifting of impacts outside system boundary
2. Keep activities within ecological capacity

Requirements for Sustainable Manufacturing

Efforts toward sustainability should

1. Prevent shifting of impacts outside system boundary
2. Keep activities within ecological capacity

- These are necessary but not sufficient conditions for sustainability

Prevent Shifting of Impacts

- Technologies meant to be environmentally friendlier may result in unintended harm
- Commonly caused due to shifting of impacts outside system boundary
  - Environmental impacts can shift across space, time, disciplines, flows
  - Examples: chlorofluoro carbon compounds, biofuels, lighting
- Current methods mainly consider shifting across spatial scales and multiple flows
  - Life cycle assessment (LCA) and footprint methods

Stay Within Ecological Capacity

- Demand for ecosystem services should not exceed supply
- Exceeding ecological capacity causes ecological degradation
- Needs to be considered at multiple spatial scales

Motivation Stay within Ecological Capacity

Stay Within Ecological Capacity

- Demand for ecosystem services should not exceed supply
- Exceeding ecological capacity causes ecological degradation
- Needs to be considered at multiple spatial scales

Urban Air Pollution

http://time.com/4259548/mexico-city-air-pollution-alert/

Sustainable Engineering 4 / 19

Stay Within Ecological Capacity

- Demand for ecosystem services should not exceed supply
- Exceeding ecological capacity causes ecological degradation
- Needs to be considered at multiple spatial scales

Shrinking of Aral Sea

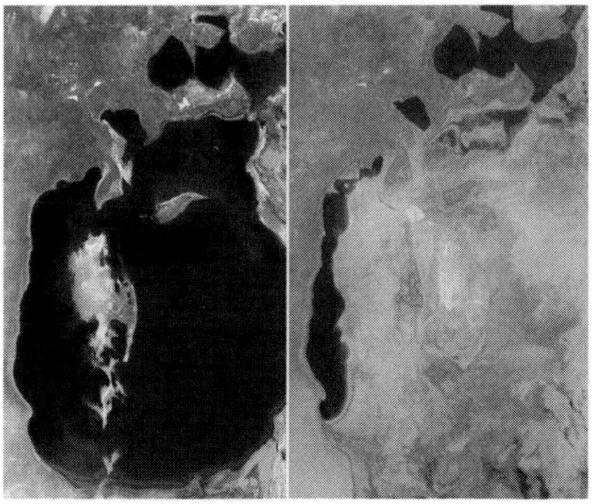

https://commons.wikimedia.org/wiki/File:AralSea1989_2014.jpg

Motivation Stay within Ecological Capacity

Stay Within Ecological Capacity

- Demand for ecosystem services should not exceed supply
- Exceeding ecological capacity causes ecological degradation
- Needs to be considered at multiple spatial scales

Lake Erie Algae

http://www.nytimes.com/interactive/2013/03/15/us/algae_in_lake_erie.html

Stay Within Ecological Capacity

- Demand for ecosystem services should not exceed supply
- Exceeding ecological capacity causes ecological degradation
- Needs to be considered at multiple spatial scales

CO_2 Measurements

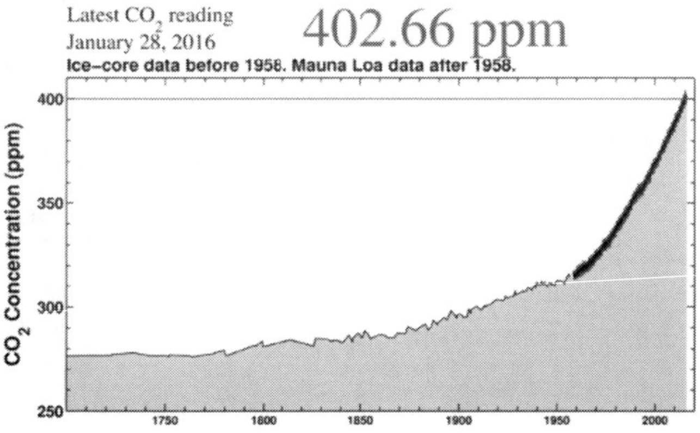

http://keelingcurve.ucsd.edu/

Motivation Stay within Ecological Capacity

Stay Within Ecological Capacity

- Demand for ecosystem services should not exceed supply
- Exceeding ecological capacity causes ecological degradation
- Needs to be considered at multiple spatial scales
- Current methods focus on ecological impacts but ignore capacity

Planetary Boundaries

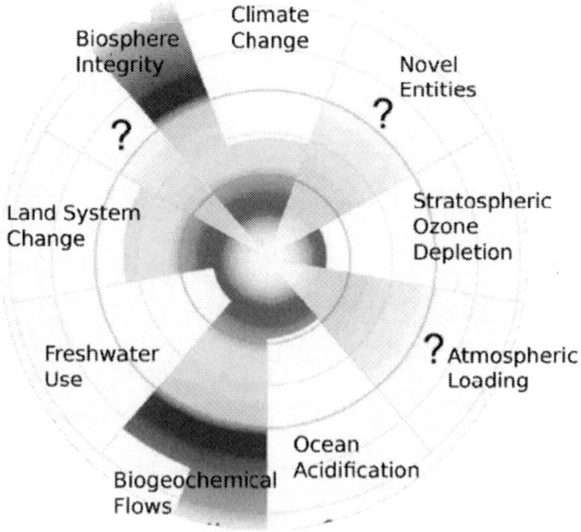

Steffen et al., *Science*, 2015

Sustainable Engineering 4 / 19

Objectives

Develop framework for including nature in engineering

- Assess nature's capacity to satisfy industrial needs
- Design integrated networks of technological and ecological systems
- Consider systems at multiple spatial scales - from individual process to their life cycles

Ecosystem Goods and Services

- Ecosystems provide goods and services that are essential for human well-being, but most methods ignore ecological capacity

Ecosystem Goods and Services

- Ecosystems provide goods and services that are essential for human well-being, but most methods ignore ecological capacity

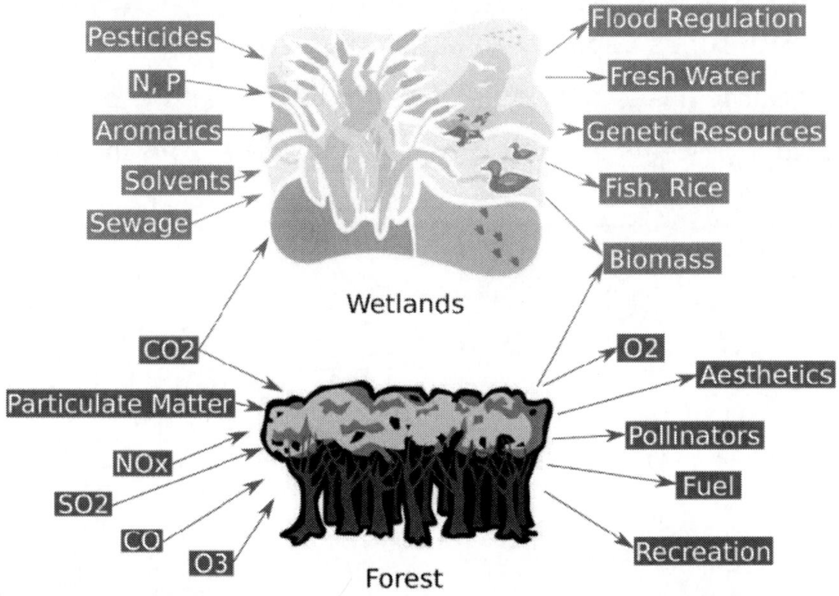

Techno-Ecological Synergy

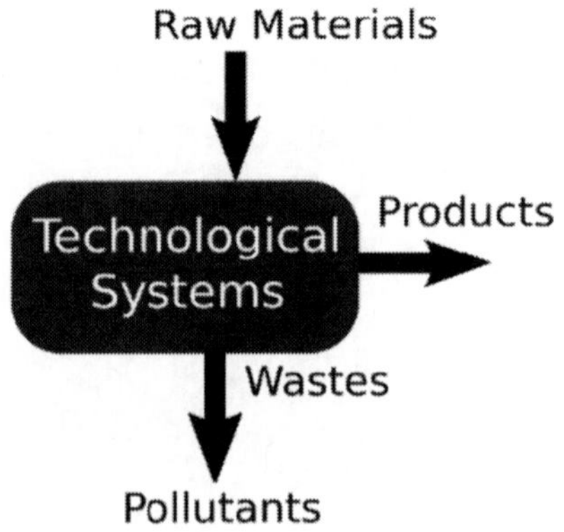

- Eco-efficiency, life cycle design

Bakshi, Ziv, Lepech, *Env. Sci. Technol.*, 2015

Techno-Ecological Synergy

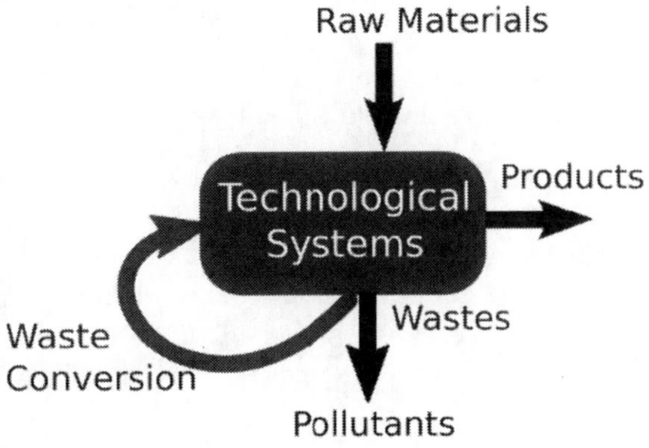

- Eco-efficiency, life cycle design
- Circular economy, industrial symbiosis, byproduct synergy

Bakshi, Ziv, Lepech, *Env. Sci. Technol.*, 2015

Techno-Ecological Synergy

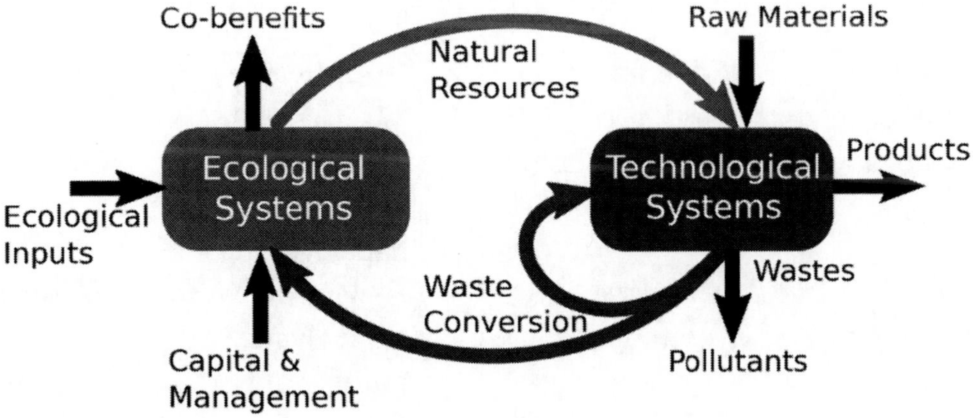

- Eco-efficiency, life cycle design
- Circular economy, industrial symbiosis, byproduct synergy
- Techno-ecological synergy

Bakshi, Ziv, Lepech, *Env. Sci. Technol.*, 2015

Techno-Ecological Synergy

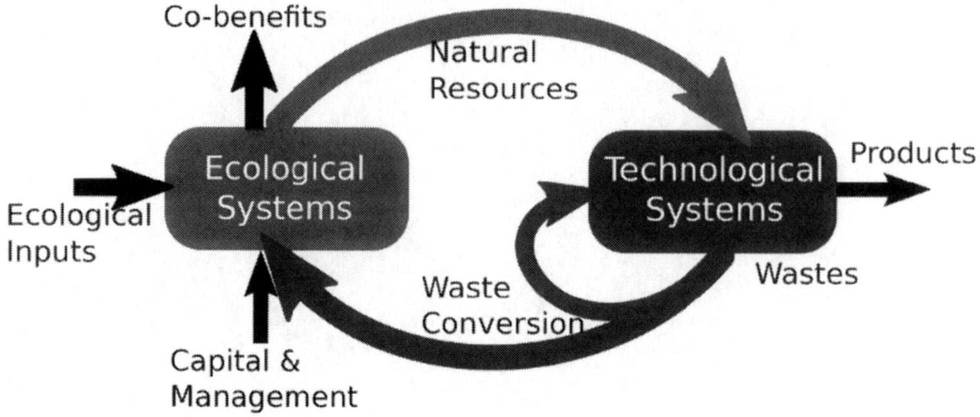

- Eco-efficiency, life cycle design
- Circular economy, industrial symbiosis, byproduct synergy
- Techno-ecological synergy
- Sustainable TES

Bakshi, Ziv, Lepech, *Env. Sci. Technol.*, 2015

Ecosystem Services: Demand and Supply

- Quantify demand, $D_{i,j,k}$ and supply, $S_{i,j,k}$ of ecosystem services

Ecosystem Service, k	**Demand, $D_{i,j,k}$**	**Supply, $S_{i,j,k}$**	**Largest Scale, J**
C sequestration	CO_2 emission	Ecological sequestration capacity	Global
Water provisioning	Water withdrawal	Renewable water from rivers, aquifers, etc.	Watershed
Air quality regulation	Air pollutants	Cleaning capacity of trees, wind, etc.	Regional
Pollination	Flowers needing pollination	Availability of pollinators	Local

TES of Corn Ethanol Life Cycles

- Consider technological and ecological systems at multiple spatial scales
- Sustainability requires that for each ecosystem service,
  - Supply $\geq$ Demand
- For global sustainability, condition needs to be satisfied at scale of serviceshed

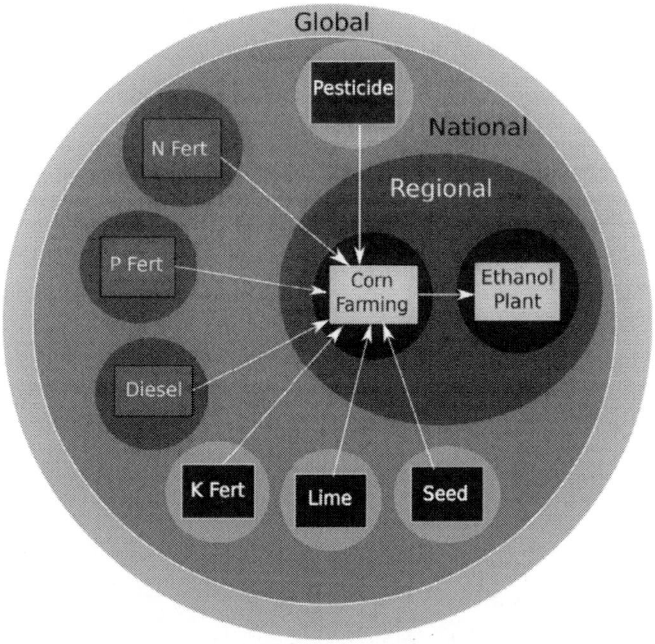

TES + LCA: Till versus No Till Farming

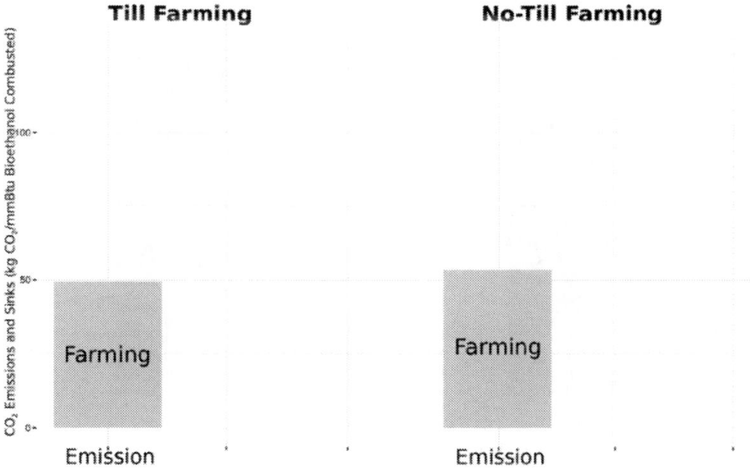

TES + LCA: Till versus No Till Farming

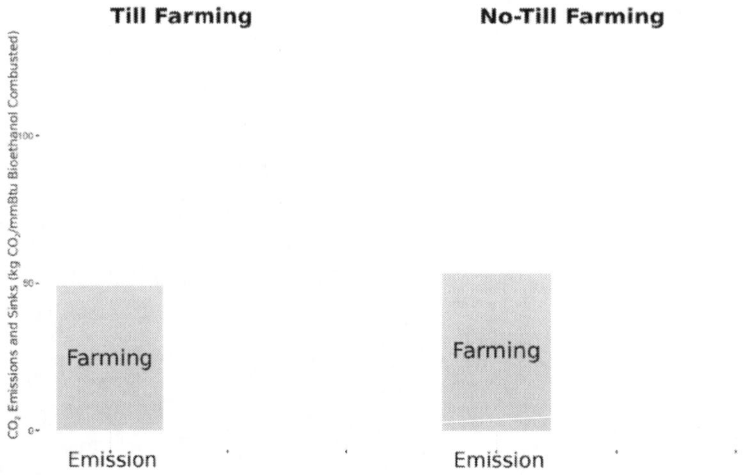

- LCA considers only demand (emissions): Till farming is better

TES + LCA: Till versus No Till Farming

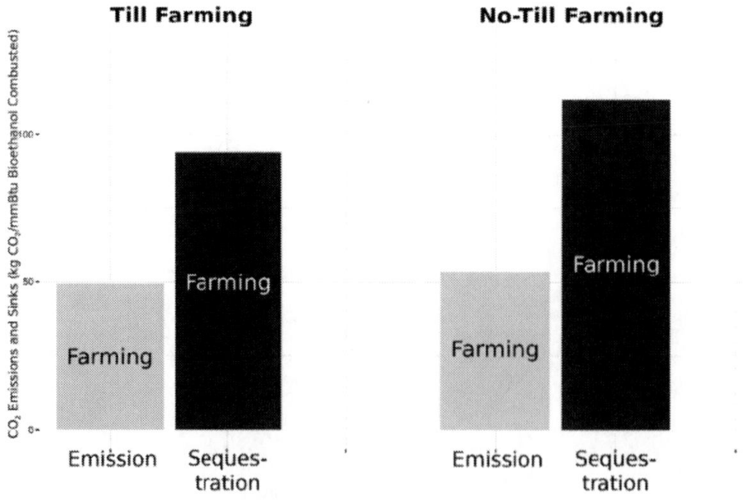

- LCA considers only demand (emissions): Till farming is better

TES + LCA: Till versus No Till Farming

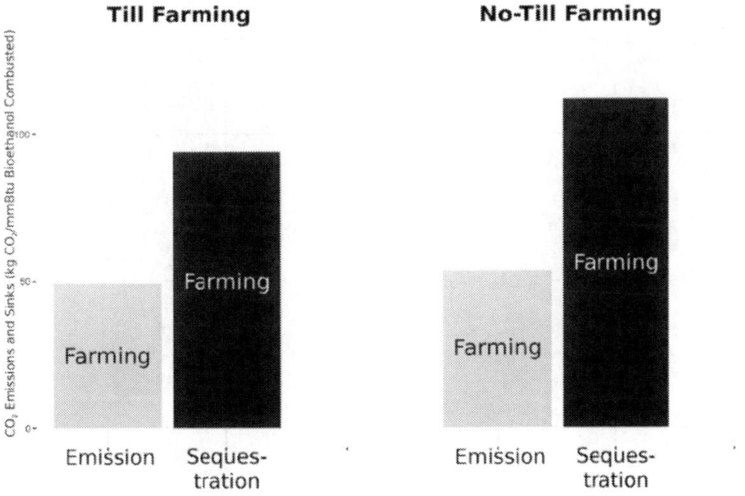

- LCA considers only demand (emissions): Till farming is better
- TES considers demand and supply: No-till farming is better and locally sustainable

TES + LCA: Till versus No Till Farming

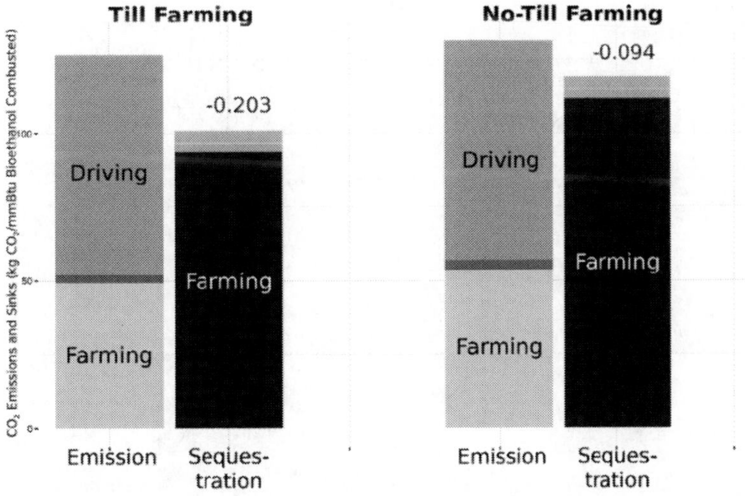

- LCA considers only demand (emissions): Till farming is better
- TES considers demand and supply: No-till farming is better and locally sustainable

TES + LCA: Till versus No Till Farming

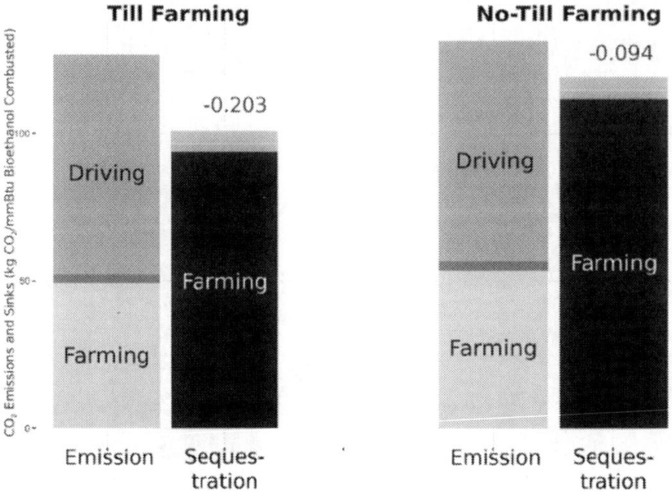

- LCA considers only demand (emissions): Till farming is better
- TES considers demand and supply: No-till farming is better and locally sustainable
- Full life cycle is locally unsustainable, but no-till is better

Assessing TES: Biodiesel Manufacturing Site

Conventional Flowsheet: Biodiesel Manufacture

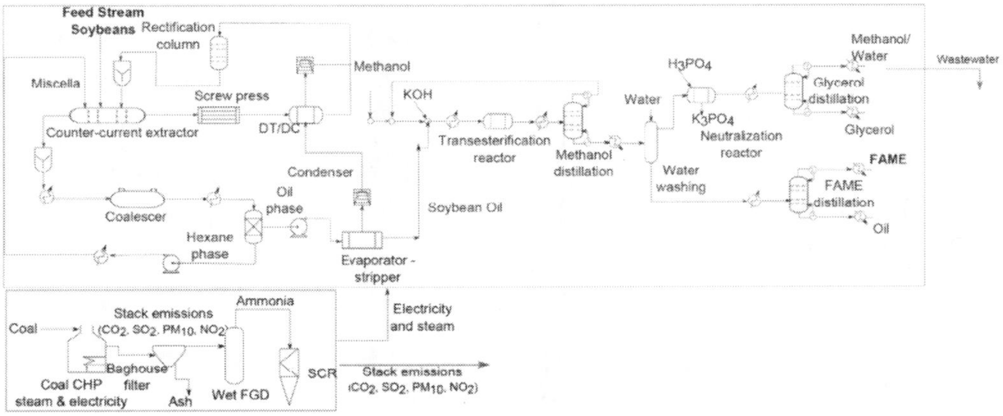

Emissions from Biodiesel Process

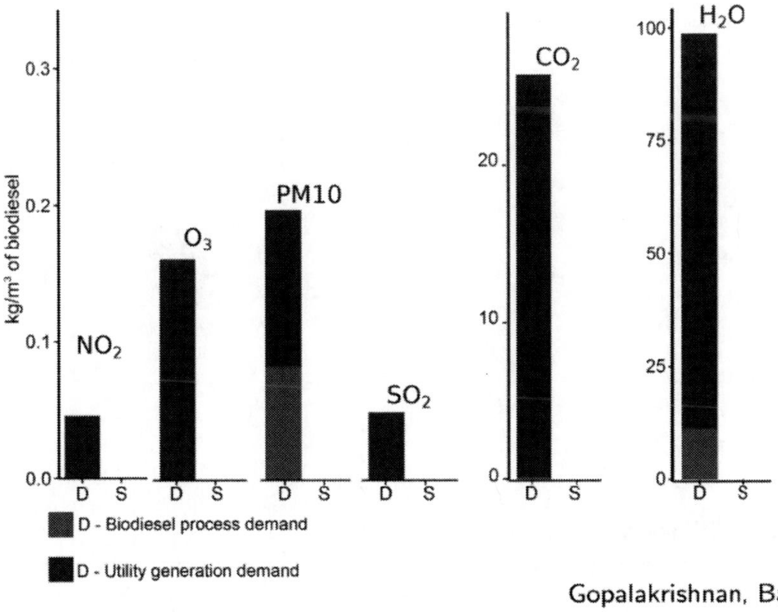

Gopalakrishnan, Bakshi, Ziv, *AIChE J.*, 2016

Emissions from Biodiesel Process

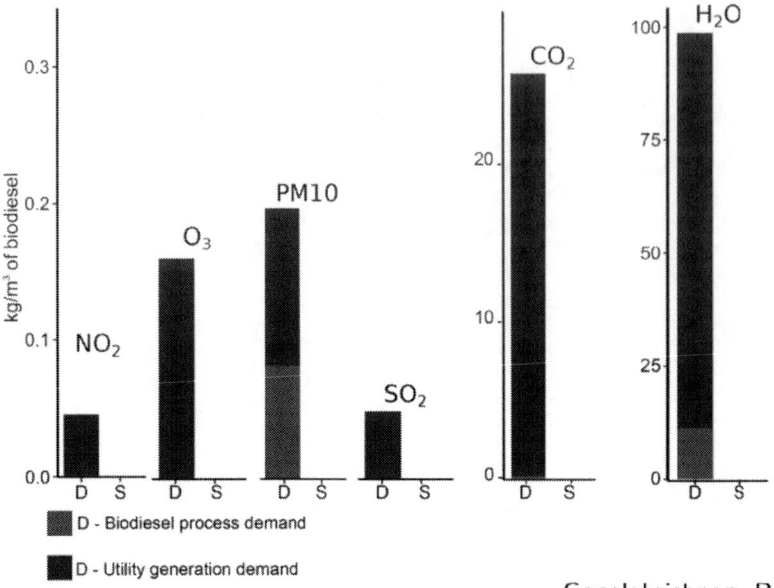

- Emssions and resource use quantify demand for ecosystem services
- Traditional efforts modify process to reduce emissions
- Emphasis on doing "less bad"

Gopalakrishnan, Bakshi, Ziv, *AIChE J.*, 2016

Biodiesel Process with Ecosystems as Unit Operations

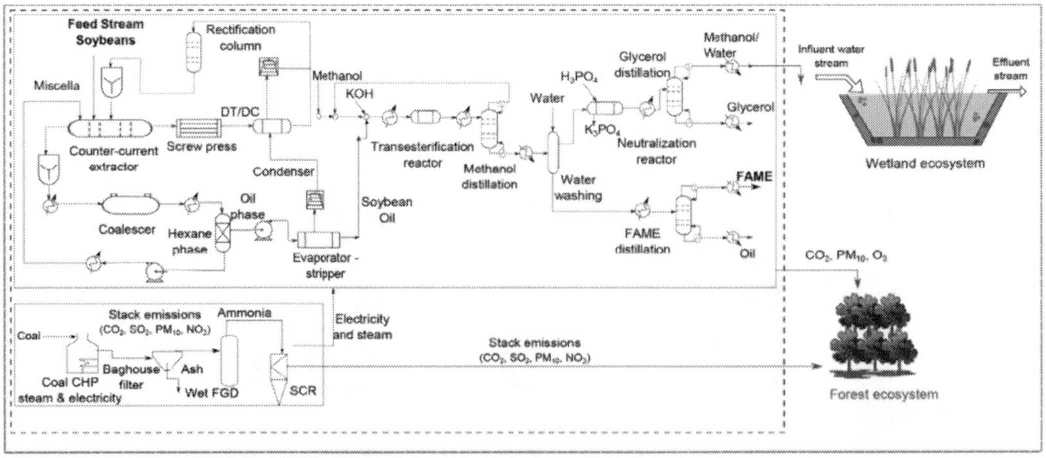

- Goal is to achieve zero net emissions from site
- Environmental regulations are already satisfied

Biodiesel TES with Forest and Wetland

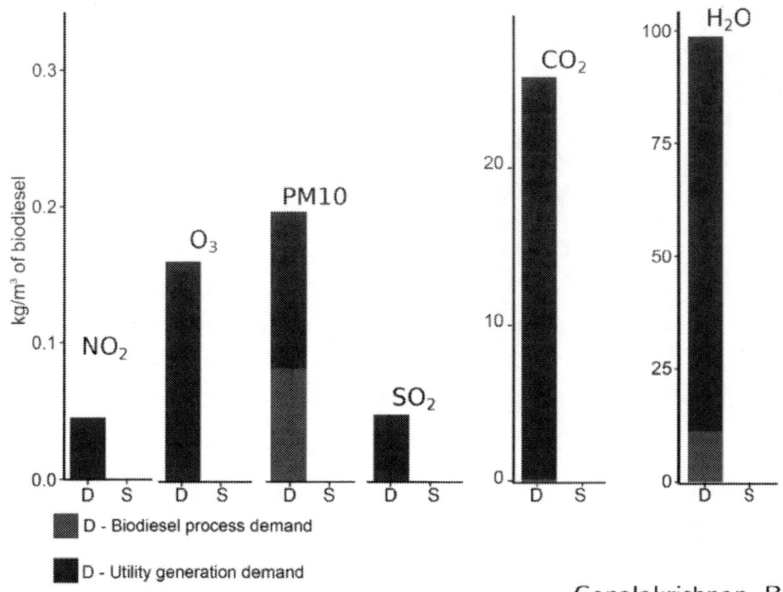

Gopalakrishnan, Bakshi, Ziv, *AIChE J.*, 2016

Synergies with Nature TES

Biodiesel TES with Forest and Wetland

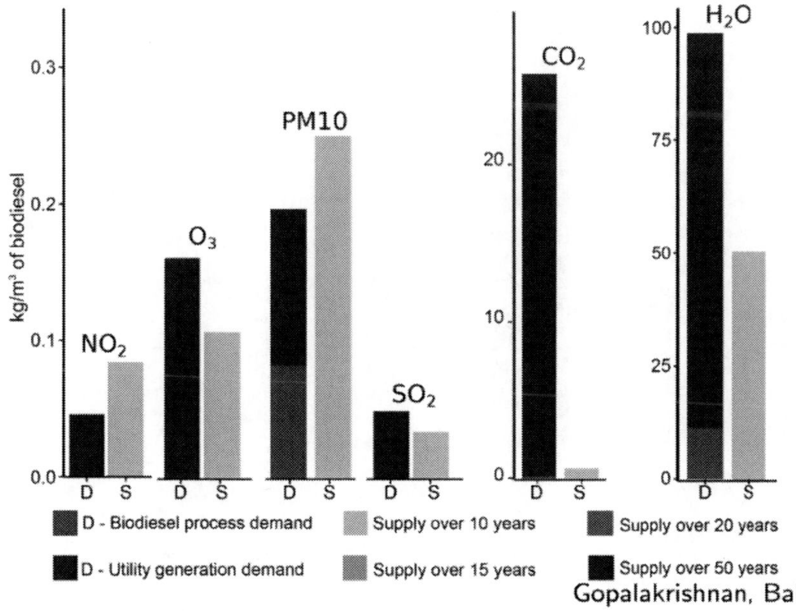

- 10 year-old forest can capture all NO_2 and PM10

Gopalakrishnan, Bakshi, Ziv, *AIChE J.*, 2016

Biodiesel TES with Forest and Wetland

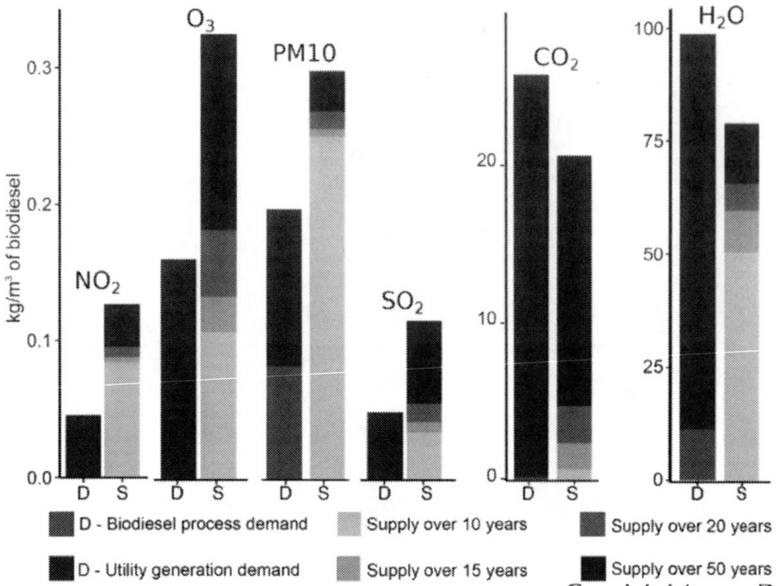

- 10 year-old forest can capture all NO_2 and PM10
- 20 year-old forest captures all air emissions except CO_2

Gopalakrishnan, Bakshi, Ziv, *AIChE J.*, 2016

Biodiesel TES with Forest and Wetland

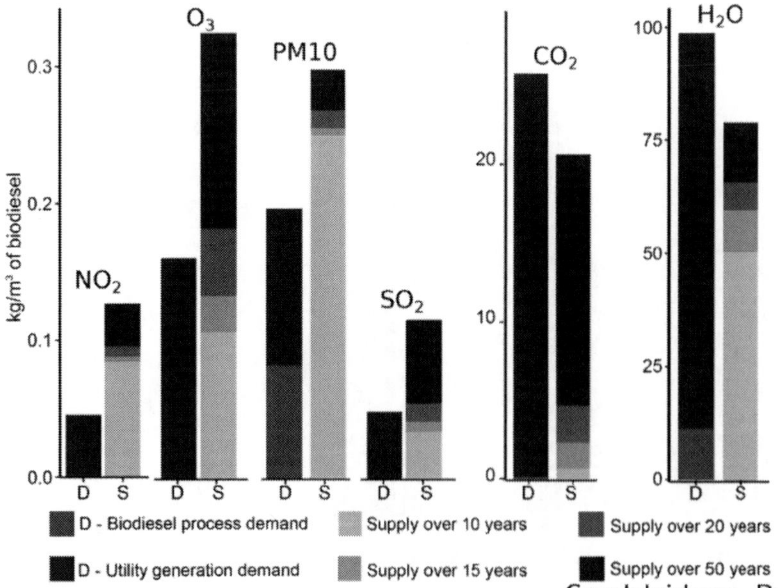

- 10 year-old forest can capture all NO_2 and PM10
- 20 year-old forest captures all air emissions except CO_2
- 50 year-old forest mitigates 80% CO_2 and H_2O

Gopalakrishnan, Bakshi, Ziv, *AIChE J.*, 2016

Biodiesel TES: Cost Analysis

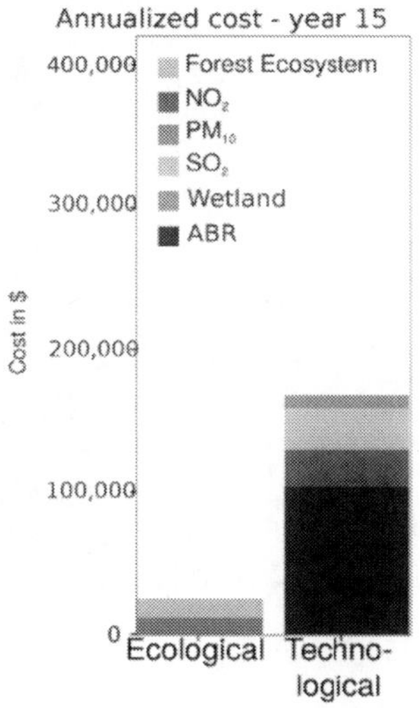

Biodiesel TES: Cost Analysis

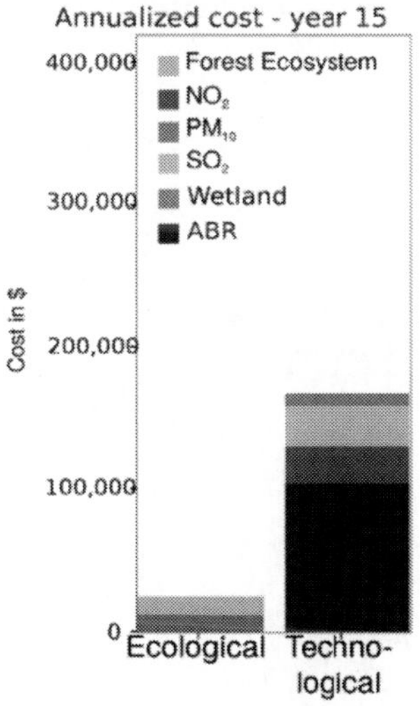

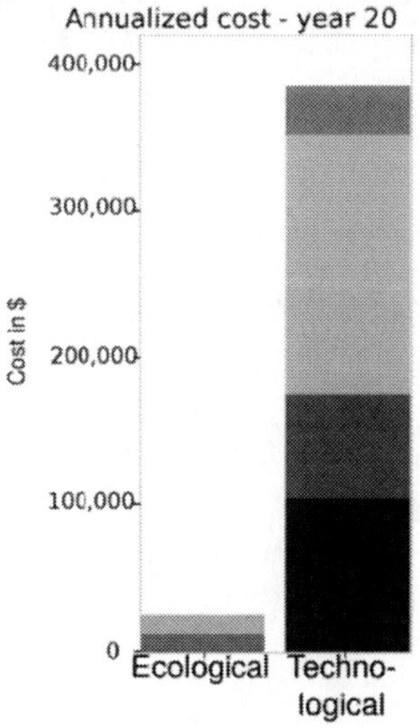

Biodiesel TES: Cost Analysis

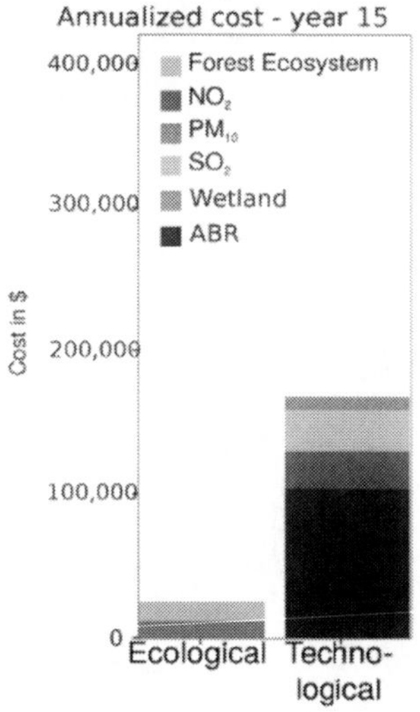

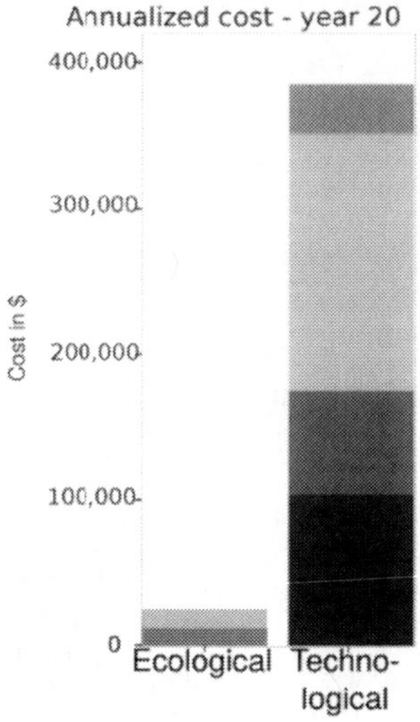

- Ecological solution is less expensive than technological solutions
- Technology depreciates over time, while ecosystems appreciate
- Ecosystems provide many additional co-benefits
- TES also gives higher ROI than conventional flowsheet

TES Design at Local Scale - Findings

Diverse applications

- Biodiesel manufacturing
- Biosolids management in Central Ohio
- Single-family home and yard
- Agricultural landscape design

TES Design at Local Scale - Findings

Diverse applications

- Biodiesel manufacturing
- Biosolids management in Central Ohio
- Single-family home and yard
- Agricultural landscape design

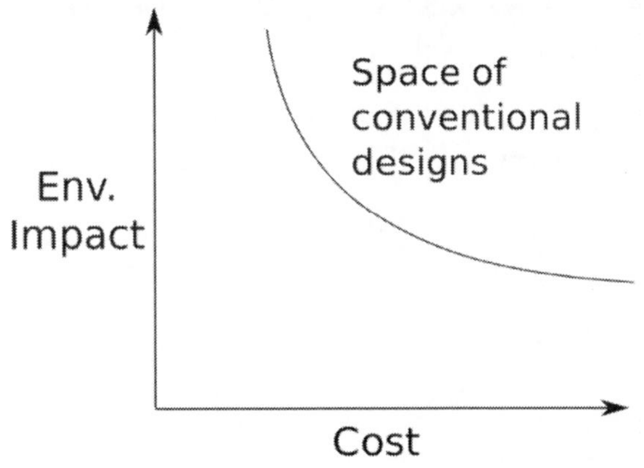

TES Design at Local Scale - Findings

Diverse applications

- Biodiesel manufacturing
- Biosolids management in Central Ohio
- Single-family home and yard
- Agricultural landscape design

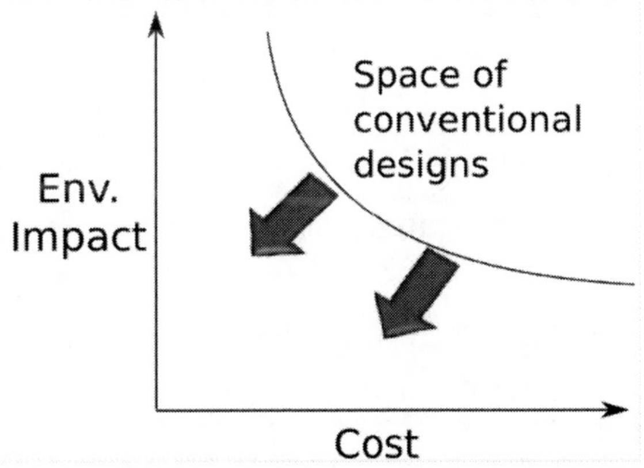

TES Design at Local Scale - Findings

Diverse applications

- Biodiesel manufacturing
- Biosolids management in Central Ohio
- Single-family home and yard
- Agricultural landscape design

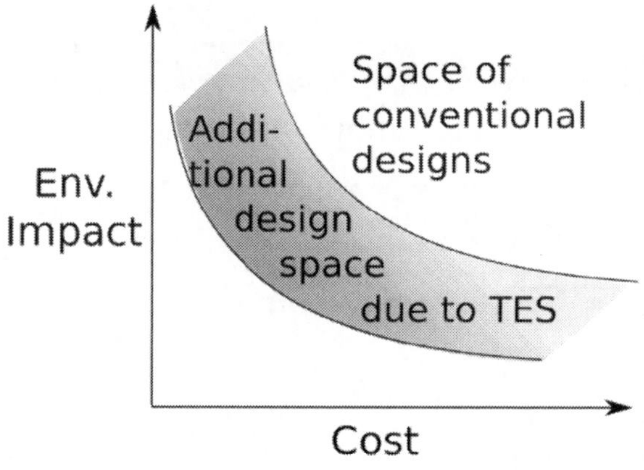

Innovation by TES

TES Design at Local Scale - Findings

Diverse applications

- Biodiesel manufacturing
- Biosolids management in Central Ohio
- Single-family home and yard
- Agricultural landscape design

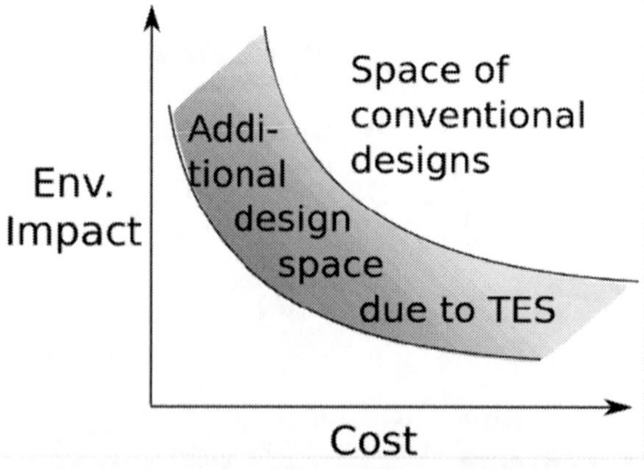

Benefits of including nature in design

- Discovers innovative designs by expanding the design space
- New designs can be "win-win"
- Need new tools and methods to realize these benefits

Sustainable Engineering 17 / 19

Summary

- Engineering has taken nature for granted, resulting in
  - Ecological degradation and unintended harm
  - Lost opportunities for economically and ecologically beneficial innovation

Summary

- Engineering has taken nature for granted, resulting in
  - Ecological degradation and unintended harm
  - Lost opportunities for economically and ecologically beneficial innovation
- Techno-Ecological Synergy
  - Assesses and encourages mutually beneficial synergies between human activities and ecosystems
  - Can provide innovative solutions that are economically and environmentally superior to conventional designs
  - Toward net positive impact

Summary

- Engineering has taken nature for granted, resulting in
  - Ecological degradation and unintended harm
  - Lost opportunities for economically and ecologically beneficial innovation
- Techno-Ecological Synergy
  - Assesses and encourages mutually beneficial synergies between human activities and ecosystems
  - Can provide innovative solutions that are economically and environmentally superior to conventional designs
  - Toward net positive impact
- Many research challenges that can benefit from industrial collaboration

Acknowledgments

- Students & Post-Docs
  - Erin Gibbemeyer
  - Varsha Gopalakrishnan
  - Rebecca Hanes
  - Xinyu Liu
  - Shweta Singh
  - Bob Urban
- Collaborators
  - Tim Gutowski, MIT
  - Satoshi Hirabayashi, USDA Forest Service
  - Michael Lepech, Stanford
  - Dusan Sekulic, U. Kentucky
  - Guy Ziv, U. Leeds
- Support
  - National Science Foundation
  - U.S. Department of Agriculture
  - U.S. Environmental Protection Agency
  - Eastman Chemical
  - American Electric Power
  - OSU Center for Resilience

Reducing Water Usage in Bioprocessing

C. Siletti
Intelligen, Inc.

Intelligen, Inc.

Biotech Water Usage

- Fermentation/Cell Culture → Purification (chromatography & membrane filtration)
- Growth Media
- Buffers
- Dilution & rinse water
- Cleaning solutions and rinses used during equipment clean-in-place (CIP).

Batch Simulation

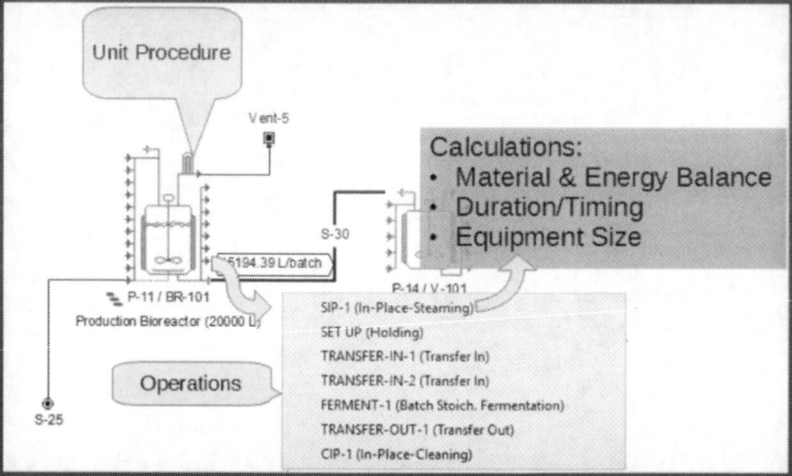

Outputs

- **M&E Balances of Integrated Processes**
- **Equipment Sizing and Costing**
- **Project Economic Evaluation**
- **Process Scheduling & Cycle Time & Throughput**

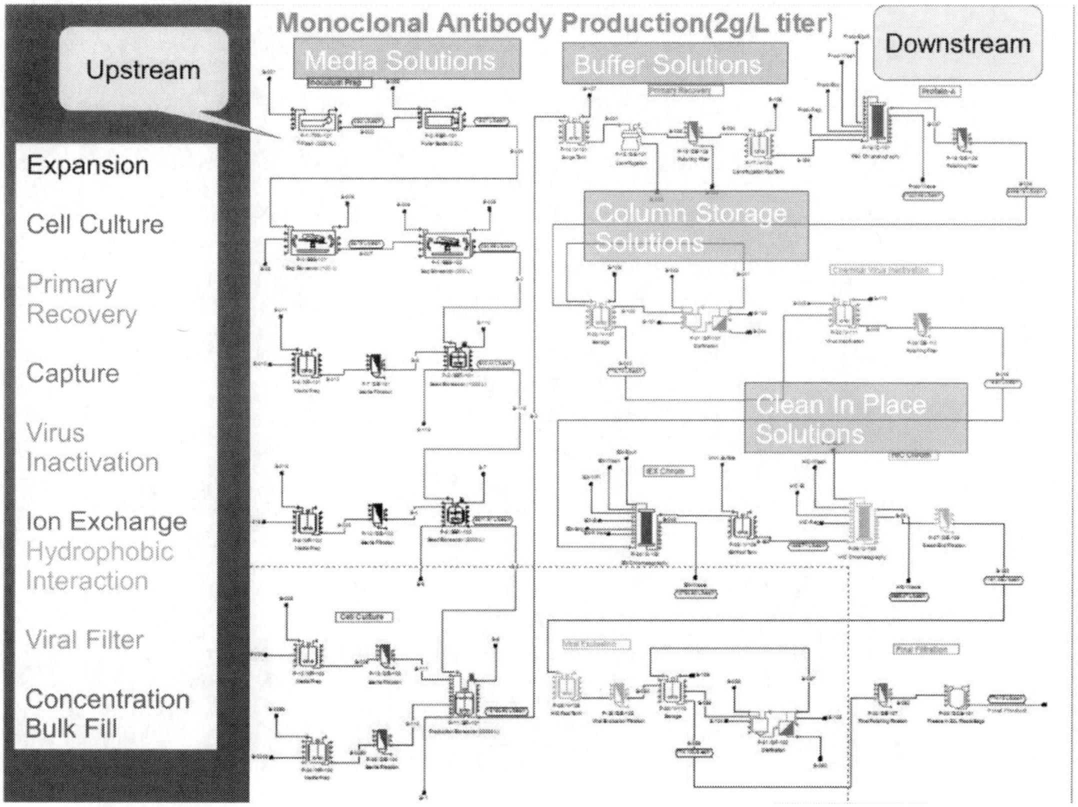

Study Case

- 15, 000 L Cell culture
- 6-step expansion train: 2 flask cultures, 2 disposable bioreactors, 2 seed bioreactors, 1 production bioreactor
- Centrifugation
- Capture, VI, Ion Exchange, HIC
- Product yield = 20kg/batch.
- Water usage 336,000L/batch or 16.8L/g product

Water Usage Distribution

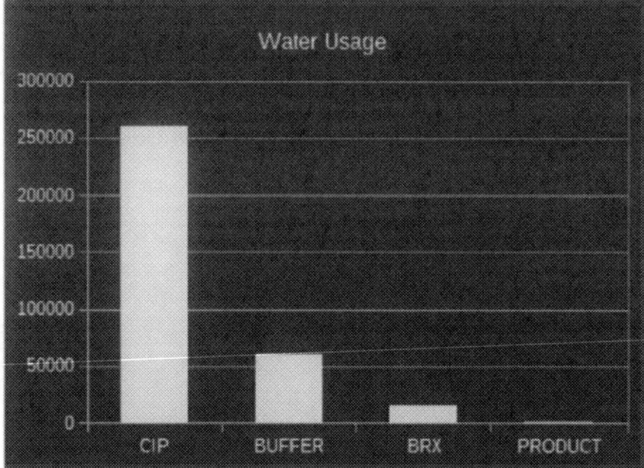

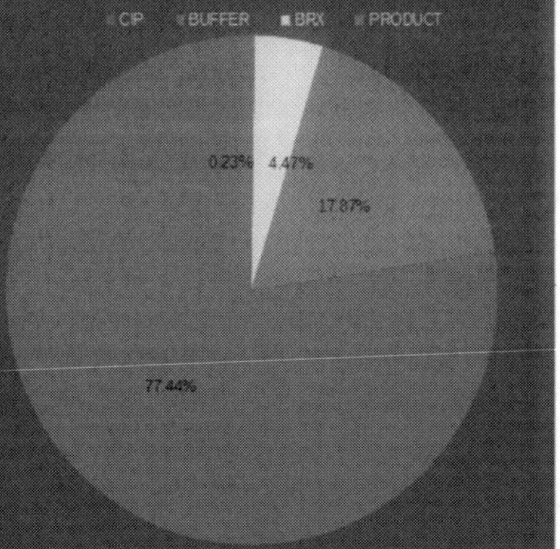

Clean In Place

Pre-rinse
Caustic
Rinse
Acid wash
Final Rinse

5-30 minute steps

Flow scales with vessel internal area

Reduction Strategies

- Eliminate the vessel (single use technology/alternative)
- Reduce vessel size
- Modify the process

CIP Usage in the Process

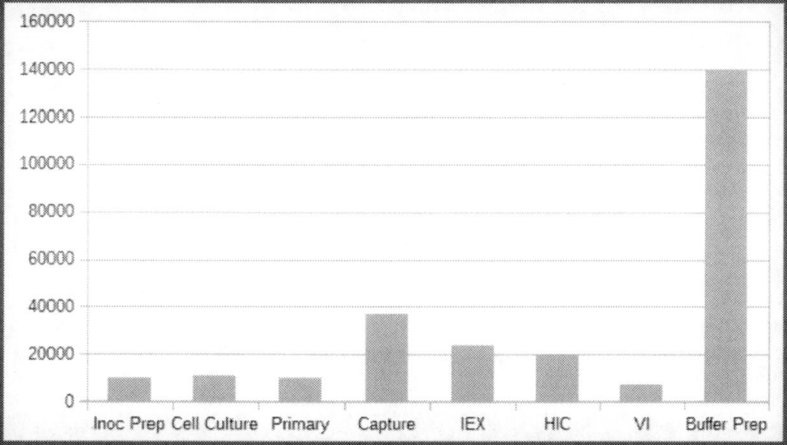

Scales with Bioreactor Size   Scales with Protein Loading

Single Use Vessels for All Vessels < 5000L

Media & Buffer Prep and Hold < 5kL
121000 L/batch 13.5% reduction

Economic Note:

Base Case Unit Production Cost	Single Use Unit Production Cost
$93/g	$91/g

Savings in labor/fixed equipment balance cost of disposables

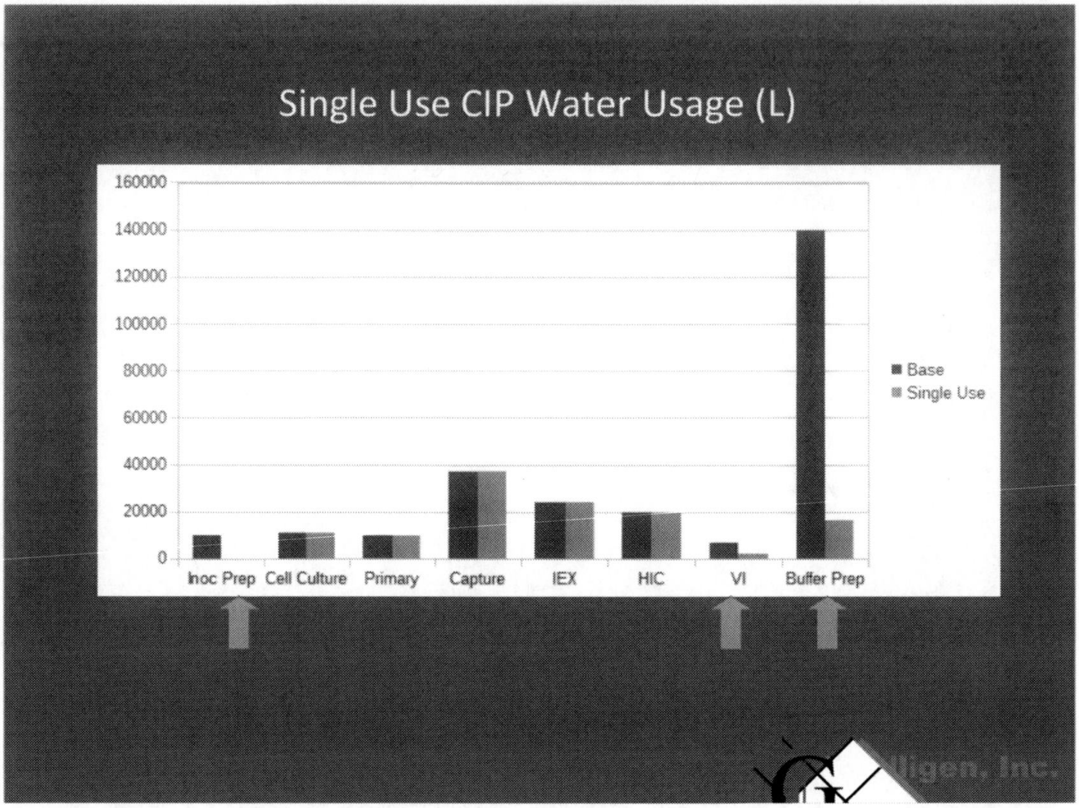
Single Use CIP Water Usage (L)
160000
140000
120000
100000
80000
60000
40000
20000
0
Inoc Prep
Cell Culture
Primary
Capture
IEX
HIC
VI
Buffer Prep
Base
Single Use

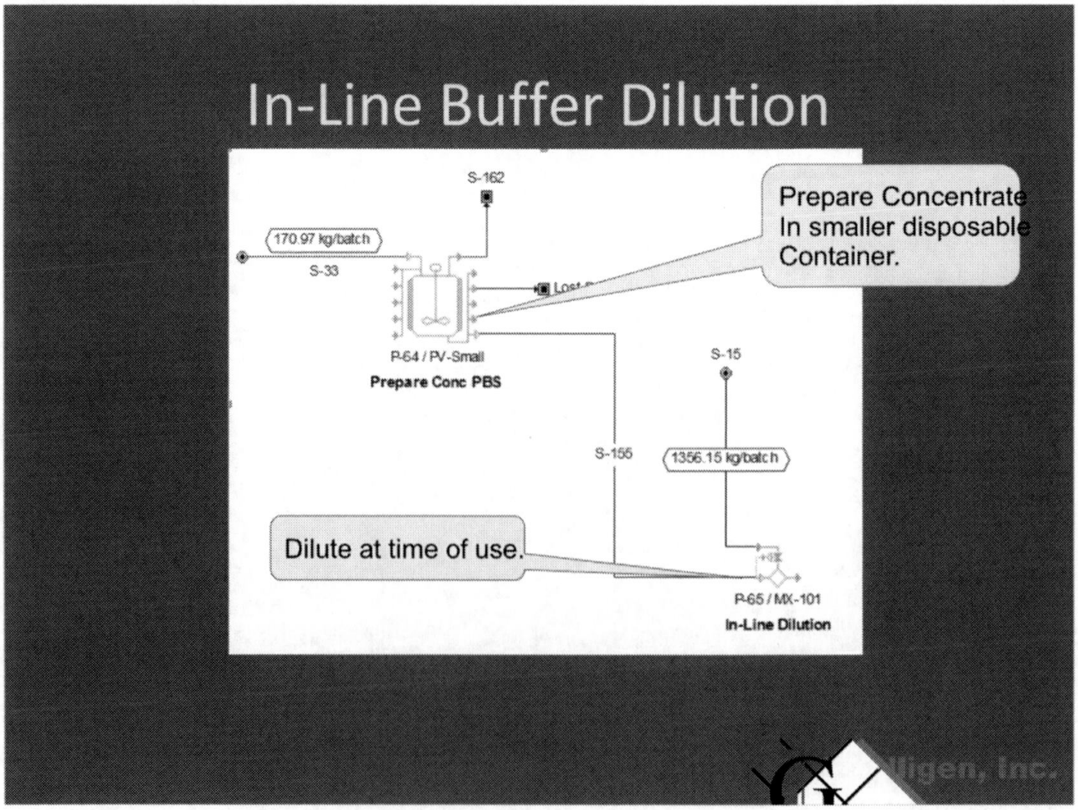
In-Line Buffer Dilution
S-162
170.97 kg/batch
S-33
Prepare Concentrate
In smaller disposable
Container.
P-64 / PV-Small
Prepare Conc PBS
S-15
S-155
1356.15 kg/batch
Dilute at time of use.
P-65 / MX-101
In-Line Dilution

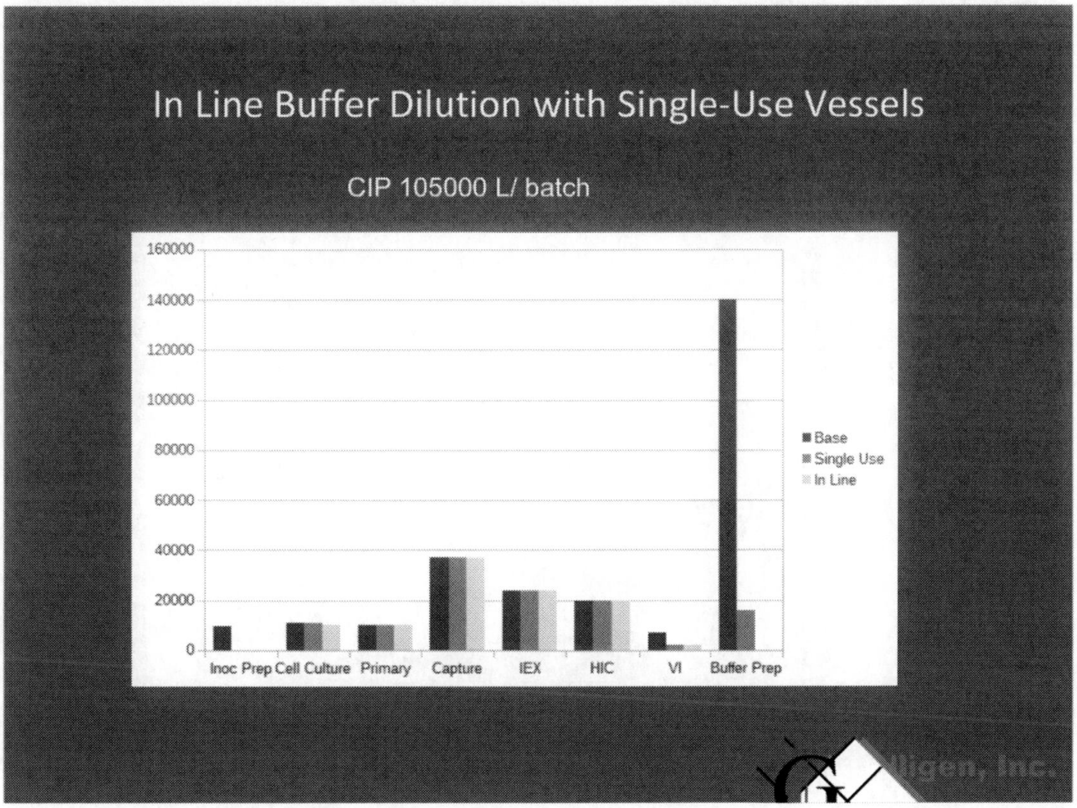
In Line Buffer Dilution with Single-Use Vessels
CIP 105000 L/ batch
160000
140000
120000
100000
80000
60000
40000
20000
0
Inoc Prep
Cell Culture
Primary
Capture
IEX
HIC
VI
Buffer Prep
Base
Single Use
In Line

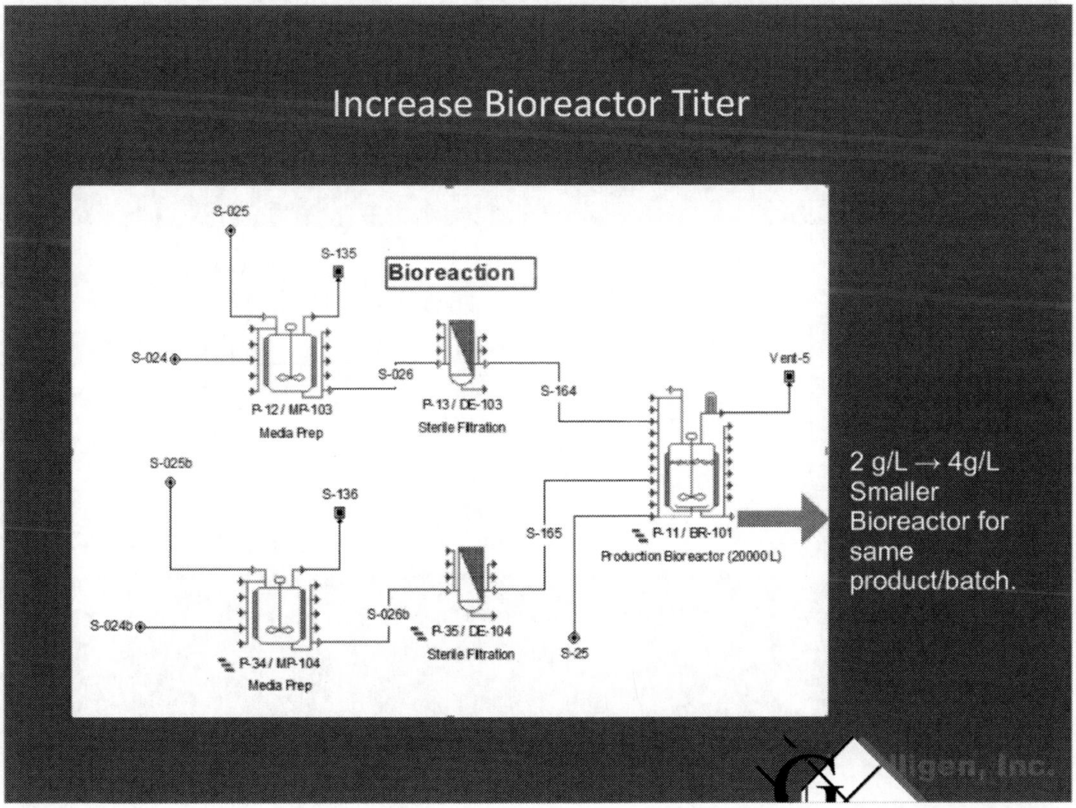
Increase Bioreactor Titer
Bioreaction
S-025
S-135
S-024
P-12 / MP-103
Media Prep
S-026
P-13 / DE-103
Sterile Filtration
S-164
Vent-5
S-025b
S-136
S-165
P-11 / BR-101
Production Bioreactor (20000 L)
S-024b
P-34 / MP-104
Media Prep
S-026b
P-35 / DE-104
Sterile Filtration
S-25
2 g/L → 4g/L
Smaller
Bioreactor for
same
product/batch.

Titer (2x) Improvement

2x Increase in Bioreactor Product Concentration
Batch Size Adjusted for 20kg/batch

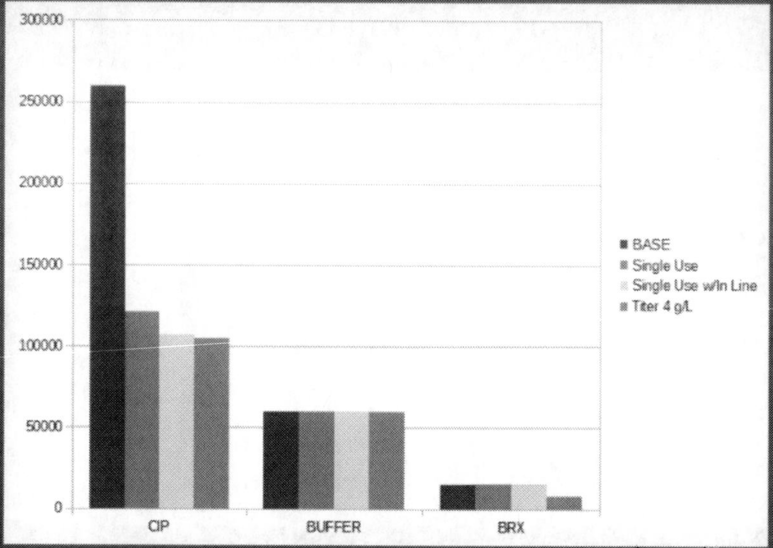

Single Use Purification Columns

Replace IEX and HIC columns with single use columns

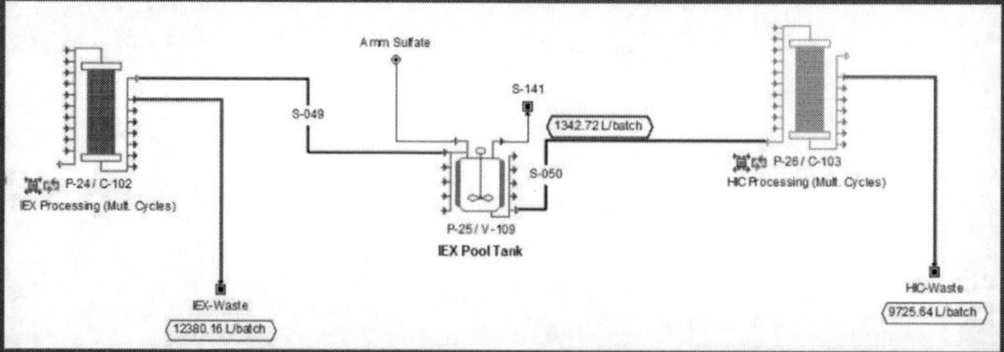

Single Use Columns

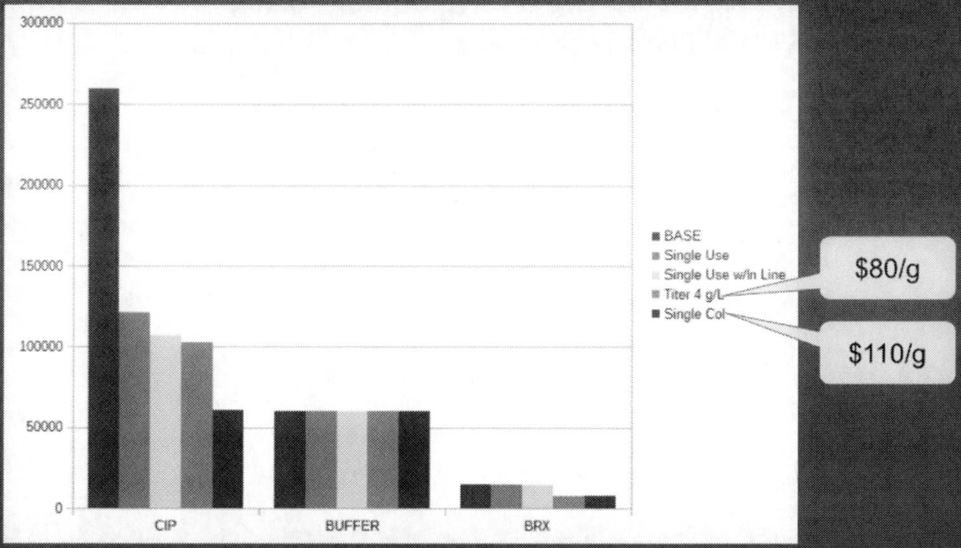

Conclusions

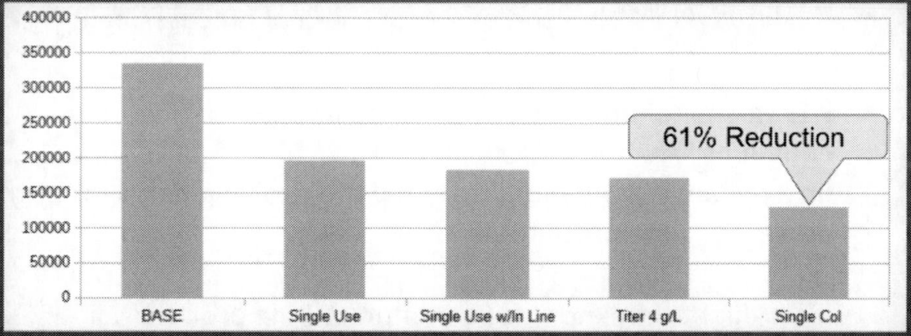

- Combination of approaches possible when different water usages are targeted
- Single use systems are the most effective, but becomes economically unattractive at large batch sizes.
- Downstream technology changes are required for maximum water reduction.

Enabling Bioprocess Development Using Virtual Plant Technology

John D. Bomberger and Tracy L. Clarke-Pringle
DuPont Engineering Research and Technology
Dynamic Simulation, Control, and Optimization Group

2016 Process Development Symposium
Providence, RI
June 9, 2016

Virtual Plant Technology

- A "Virtual Plant" is a surrogate for the real process/plant
- It is an environment where the fundamental engineering equations* that describe a process can be executed repeatedly (in a dynamic simulation), to show the evolving behavior of the process as time proceeds.
- It's a "flight simulator" for a manufacturing process

*equations: material and energy balances, thermodynamics, reaction kinetics

 2

Why Dynamic Modeling for Virtual Plants?

- Manufacturing processes, particularly those that include batch or sequential operations, are often too complex to be able to think through entire scenarios or consequences of actions

- We can't run unlimited experiments
  - equipment limitations, time limitations
  - cost of experiments
  - the future organism or equipment may not exist
  - it may not be safe

- Dynamic modeling provides a safe and educational environment to test out all ideas and gain insights into how a batch will evolve.
  - A 40 hour batch can be run in minutes

- Dynamic modeling provides quantitative data and improves our ability to make knowledge-based decisions

 3

A Competitive Advantage

- When you combine Virtual Plant Technology with good process engineers, you end up with a winning combination

- When a good engineer develops or has access to a Virtual Plant, they will learn from it:
  - Evaluate and test process alternatives and process operability
  - Evaluate ideas for optimizing the process
  - Control strategy design and testing
    - This includes both batch automation and regulatory controls
    - Addressing how abnormal situations are handled
    - Rapid prototyping for advanced process control
  - Process automation systems (DCS, PLS) configuration testing
  - Operator Training Simulator

- Doing this "virtually" is faster, cheaper, and safer than using the real process

 4

DuPont Experience with Virtual Plants: TMODS

- DuPont™ TMODS was created by DuPont engineers circa 1990. It is a first-principles, engineering, dynamic simulator environment
- Today, this DuPont-owned software is managed and supported to meet DuPont needs
  - A very small core development team
  - We are users and developers
- DuPont Engineering Research and Technology: Dynamic Simulation, Control and Optimization (DySCO) group – world class process modeling and control experts, and power users of TMODS
- The 300+ users across DuPont (process and process control engineers)
  - A wide range of power users and casual users

 5

Using DuPont™ TMODS

- DuPont™ TMODS Virtual Plants are created from a library of unit operations
- Users provide specifications for each unit op (size, # trays, UA, valve CVs, etc.)
- Models are easy to construct, and can encompass whole plants
- Interactive simulations run on business grade laptops for maximum accessibility

 6

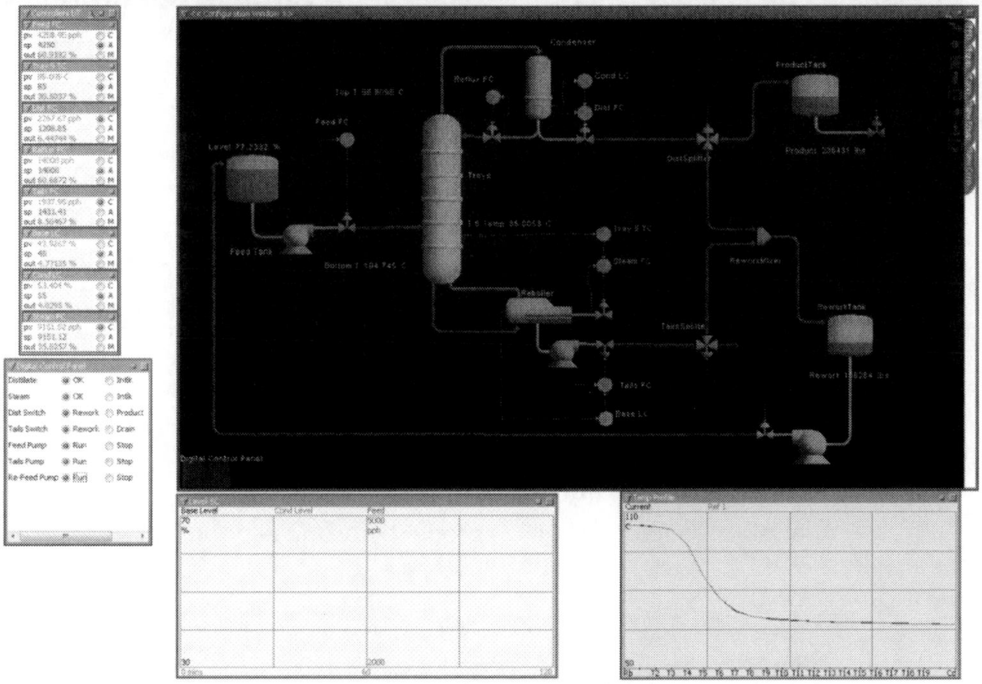

 Public

Virtual Plant Technology and Bioprocess Development

- DuPont has made a clear, long-term commitment to developing and operating bioprocesses
  - Modeling and simulation are critical in bioprocess development
    - Complicated flowsheets
    - Extensive use of batch and combined continuous / batch processing
    - Profitable in a narrow window of process conditions.
  - Dynamic models are valuable tools to develop process understanding
- Using Virtual Plant Technology in this area:
  - Accelerate growth of new businesses
  - Deliver productivity improvements

 Public

Project Life Cycle: Pre-commercialization through operation

- Example 1: Cellulosic Ethanol
  - Batch control design and functional description
  - Configuration checkout (FAT)
  - Startup procedure development and testing
  - Operator Training
- Example 2: Enzymatic Polymerization
  - Demonstration of Integrated operation
  - Source of dynamic data for individual unit operations, complementing Aspen and Excel flowsheet models
  - Plant wide modeling was started very early in the project life cycle
- Example 3: Fermentation with volatile product
  - Analysis of reactor design on productivity and economics

Example 1: Cellulosic Ethanol

- Virtual Plant Technology was used for both pilot and commercial scale design and training

- Control System Functional Descriptions
  - High level control strategy narratives written from control strategies designed and tested on DuPont™ TMODS plant models
  - Critical batch automation was designed on TMODS
  - Detailed functional description for batch automation was prepared
  - BatchML export from TMODS was used to guide system integrators while programming batch logic

Example 1: Cellulosic Ethanol

- DCS Configuration
  - DCS was *required* to be connected to DuPont™ TMODS models during Factory Acceptance Test of control system
  - Permitted thorough testing of all regulatory controls and batch automation
  - Potential time savings on Site Acceptance Test and configuration re-work
- Startup Procedures
  - Models of some areas were used to prepare and test startup procedures
  - Especially critical for highly interactive, energy integrated process

Example 1: Cellulosic Ethanol

- Operator Training
  - A DCS connected Operator Training Simulator was created
  - Uses DCS identical (in most aspects) to plant DCS
  - In OTS, DuPont™ TMODS replaces inputs from plant with inputs from model
  - DCS outputs (controller outputs, digital signals) drive TMODS model

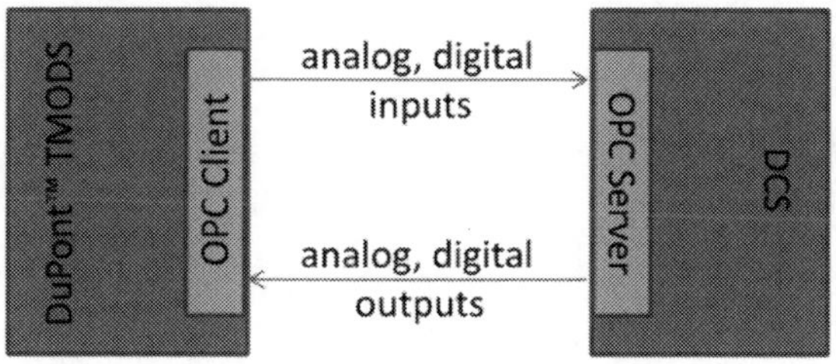

Example 2: Enzymatic Polymerization

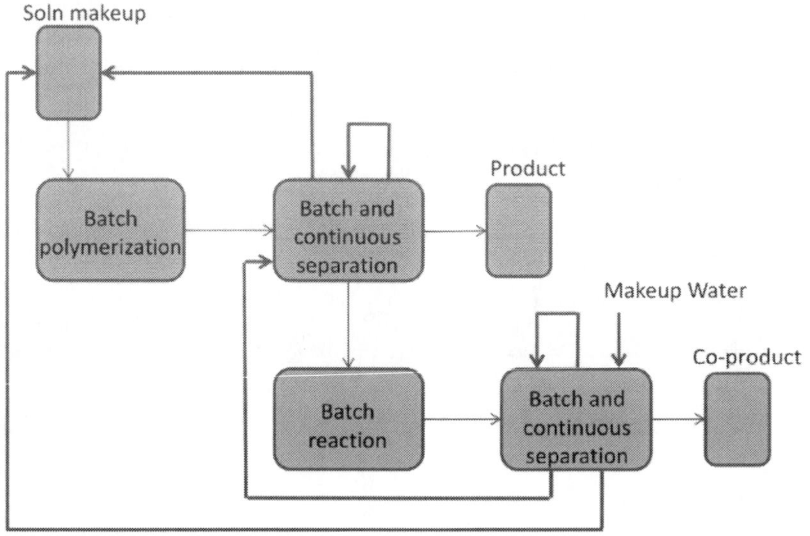

 13

Example 2: Enzymatic Polymerization

- The process design relies heavily on water reuse, recycle, and integration within and between two co-processes
  - Water recycle is necessary for process economics

- However, the piloting strategy is to demonstrate pieces of the process at a toll facility
  - The only place the full integrated process will be demonstrated is in dynamic model

 14

Example 2: Enzymatic Polymerization

- The process is a highly integrated combination of continuous, batch, and dynamically varying operations. Automation will be required to achieve design rates while minimizing labor costs
- DuPont™ TMODS provides a way to efficiently model the entire flowsheet dynamically. It is the best way we have of understanding plant-wide interactions to design and test the controls
- We have a preliminary automation design that optimizes operability of the integrated plant. Key elements:
  - Automation utilized to adjust batch, set rates based on need
  - Advanced regulatory control for soln tank inventory and concentration control
  - Surge tanks in between batch and continuous unit operations
  - Separation tanks watchdog to avoid tank drain or overflow

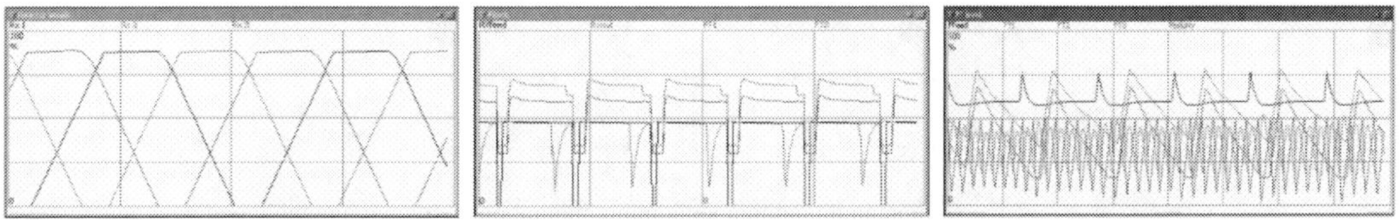

 Public 15

Example 3: Fermentation with volatile product

- What reactor type and configuration provides lowest cost (capital, variable)?
- Productivity targets (productivity, titer), production rate set
- Key problem: Aerobic fermentation product is flammable gas
- VPT was used to evaluate design and operation, and the impact on cost and capital

 Public 16

Example 3: Fermentation with volatile product

- Bubble Column versus STR?

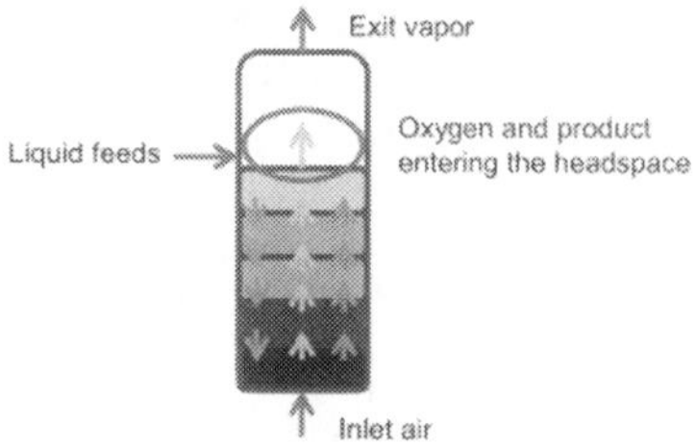

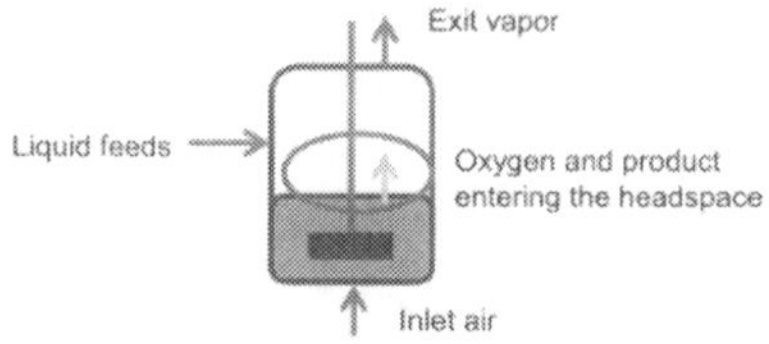

- N_2 is added to headspace to make sure mixture is not flammable
- Because of lower mass transfer in bubble column, N_2 is required throughout batch -> high variable cost than STR

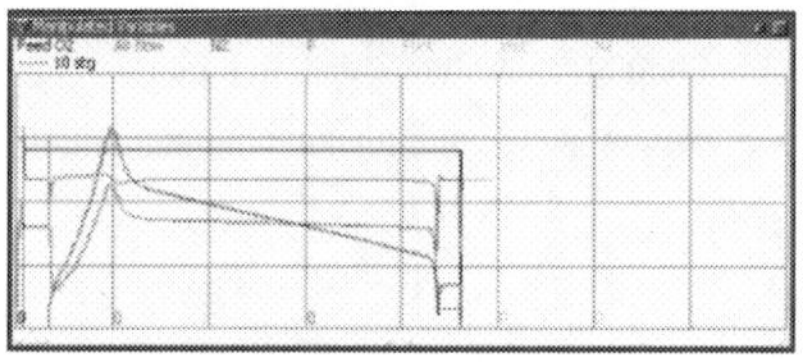

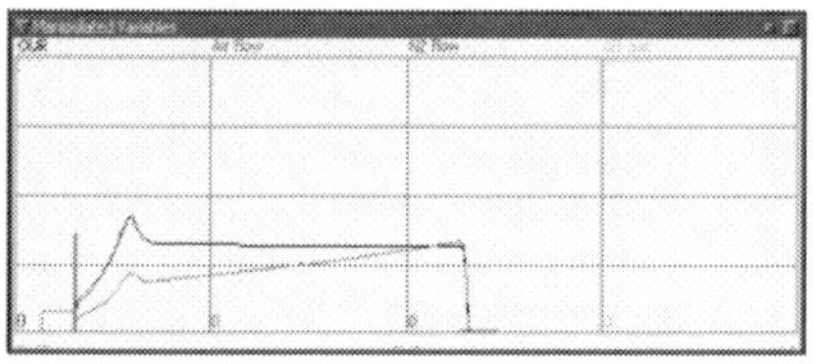

Summary

- Virtual Plant Technology is becoming an integral part of our commercialization of bioprocess technology
- It is used throughout our project lifecycle: Development, control system configuration and testing, training, and process improvement
- We find it a valuable way to save time and money during development, piloting, and commissioning

Future developments

- Smart manufacturing
- Virtual Plant as library of process and control system knowledge

Understanding Synergies between Catalytic Fast Pyrolysis and Processes Based on Conventional Energy Feedstocks through Insights from Techno-Economic Analysis (TEA) Studies

Sahir, A. H., National Renewable Energy Laboratory

Originally presented on: 6/9/2016 15:50:00 - 16:30:00

Conventional energy processes based on petroleum feedstocks (e.g. Fluid Catalytic Cracking, Hydroprocessing, Hydrocracking and Delayed Coking) can play a vital role in facilitating understanding of the process engineering aspects for in situ and ex situ catalytic fast pyrolysis vapor upgrading. The higher oxygen content of biomass and its tendency towards coke and gaseous species formation serves as a major differentiating factor between conventional petroleum refining processes, and biomass-based thermochemical processes like pyrolysis.

The development of 21 billion gallons of advanced biofuels annually by 2022 is mandated by the U.S. Energy Security and Independence Act of 2007. The potential of in situ and ex situ upgrading of fast pyrolysis vapors as one of the promising research options for thermochemical conversion of biomass to liquid transportation fuels has been explored in a thermochemical research pathway process design report (http://www.nrel.gov/docs/fy15osti/62455.pdf). Based on the understanding derived from process engineering literature on Fluid Catalytic Cracking, the possible implications of a significant coke content observed in current experimental studies on catalytic fast pyrolysis vapor upgrading can be understood, and will be discussed in this presentation.

AUTHOR INDEX

AIChE
120 Wall Street, FL 23
New York, NY 10005-4020

ISBN 978-1-5108-2576-5

9 781510 825765

CURRAN ASSOCIATES INC.
proceedings.com